Lectures in Applied Mathematics
Volumes in This Series

Proceedings of the Summer Seminar, Boulder, Colorado, 1960

1 **Lectures in Statistical Mechanics,** *G. E. Uhlenbeck and G .W. Ford with E. W. Montroll*
2 **Mathematical Problems of Relativistic Physics,** *I. E. Segal with G. W. Mackey*
3 **Perturbation of Spectra in Hilbert Space,** *K. O. Friedrichs*
4 **Quantum Mechanics,** *R. Jost*

Proceedings of the Summer Seminar, Ithaca, New York, 1963

5 **Space Mathematics. Part 1,** *J. Barkley Rosser, Editor*
6 **Space Mathematics. Part 2,** *J. Barkley Rosser, Editor*
7 **Space Mathematics. Part 3,** *J. Barkley Rosser, Editor*

Proceedings of the Summer Seminar, Ithaca, New York, 1965

8 **Relativity Theory and Astrophysics. 1. Relativity and Cosmology,** *Jürgen Ehlers, Editor*
9 **Relativity Theory and Astrophysics. 2. Galactic Structure,** *Jürgen Ehlers, Editor*
10 **Relativity Theory and Astrophysics. 3. Stellar Structure,** *Jürgen Ehlers, Editor*

Proceedings of the Summer Seminar, Stanford, California, 1967

11 **Mathematics of the Decision Sciences, Part 1,** *George B. Dantzig and Arthur F. Veinott, Jr., Editors*
12 **Mathematics of the Decision Sciences, Part 2,** *George B. Dantzig and Arthur F. Veinott, Jr., Editors*

Proceedings of the Summer Seminar, Troy, New York, 1970

13 **Mathematical Problems in the Geophysical Sciences. 1. Geophysical Fluid Dynamics,** *William H. Reid, Editor*
14 **Mathematical Problems in the Geophysical Sciences. 2. Inverse Problems, Dynamo Theory, and Tides,** *William H. Reid, Editor*

Proceedings of the Summer Seminar, Potsdam, New York, 1972

15 Nonlinear Wave Motion, *Alan C. Newell, Editor*

Proceedings of the Summer Seminar, Troy, New York, 1975

16 Modern Modeling of Continuum Phenomena, *Richard C. DiPrima, Editor*

Proceedings of the Summer Seminar, Salt Lake City, Utah, 1978

17 Nonlinear Oscillations in Biology, *Frank C. Hoppensteadt, Editor*

Proceedings of the Summer Seminar, Cambridge, Massachusetts, 1979

18 Algebraic and Geometric Methods in Linear Systems Theory, *Christopher I. Brynes and Clyde F. Martin, Editors*

Proceedings of the Summer Seminar, Salt Lake City, Utah, 1980

19 Mathematical Aspects of Physiology, *Frank C. Hoppensteadt, Editor*

Proceedings of the Summer Seminar, Chicago, Illinois, 1981

20 Fluid Dynamics in Astrophysics and Geophysics, *Norman R. Lebovitz, Editor*

Proceedings of the Summer Seminar, Chicago, Illinois, 1982

21 Applications of Group Theory in Physics and Mathematical Sciences, *Moshe Flato, Paul Sally and Gregg Zuckerman, Editors*

22 Large-Scale Computations in Fluid Mechanics, Parts 1 and 2, *B. Engquist, S. Osher and R. C. J. Somerville, Editors*

23 Nonlinear Systems of Partial Differential Equations in Applied Mathematics, Parts 1 and 2, *Basil Nicolaenko, Darryl D. Holm and James M. Hyman, Editors*

24 Reacting Flows: Combustion and Chemical Reactors, Parts 1 and 2, *G. S. S. Ludford, Editor*

Proceedings of the Summer Seminar, Minneapolis, Minnesota, 1987

25 Computational Aspects of VLSI Design with an Emphasis on Semiconductor Device Simulation, *Randolph E. Bank, Editor*

LECTURES IN APPLIED MATHEMATICS
VOLUME 25

Computational Aspects of VLSI Design with an Emphasis on Semiconductor Device Simulation

Randolph E. Bank
Editor

1990
AMERICAN MATHEMATICAL SOCIETY
PROVIDENCE, RHODE ISLAND

The Proceedings of the 1987 SIAM-AMS Summer Seminar on Computational Aspects of VLSI Design with an Emphasis on Semiconductor Device Simulation were prepared by the American Mathematical Society with support from the following sources: Air Force Office of Scientific Research, Army Research Office, and the National Science Foundation through NSF Grant DMS8611436.

1980 *Mathematics Subject Classification* (1985 *Revision*). 35G, 65L, 65M, 65N.

Library of Congress Cataloging-in-Publication Data

SIAM-AMS Summer Seminar on Computational Aspects of VLSI Design with an Emphasis on Semiconductor Device Simulation (1987: University of Minnesota, Minneapolis)

Computational aspects of VLSI design with emphasis on semiconductor device simulation/Randolph E. Bank, editor.

p. cm.–(Lectures in applied mathematics; ISSN 0075-8485; v. 25)

Proceedings of the 1987 SIAM-AMS Summer Seminar on Computational Aspects of VLSI Design with an Emphasis on Semiconductor Device Simulation was held at the University of Minnesota, Minneapolis from Apr. 30 to May 7, 1987.

Includes bibliographical references.

ISBN 0-8218-1132-0

1. Semiconductors–Simulation methods–Congresses. 2. Integrated circuits–Very large scale integration–Design and construction–Mathematical models–Congresses. I. Bank, Randolph E., 1949– . II. American Mathematical Society. III. Society for Industrial and Applied Mathematics. IV. Title. V. Series: Lectures in applied mathematics (American Mathematical Society); v. 25.

TK7871.85.S537 1987 89-18355

621.39'5–dc20 CIP

Printed in the United States of America.

This publication was typeset using $\mathcal{A}\mathcal{M}\mathcal{S}$-TeX, the American Mathematical Society's TeX macro System.

10 9 8 7 6 5 4 3 2 1 94 93 92 91 90

Contents

Preface

The eighteenth AMS-SIAM Summer Seminar on applied mathematics was held at the Institute for Mathematics and Its Applications, University of Minnesota, Minneapolis, from April 30 to May 7, 1987. The topic of the seminar was *Computational Aspects of VLSI Design with an Emphasis on Semiconductor Device Simulation.*

The organizing committee for the seminar included Randolph Bank, William Coughran, Eric Grosse, Mitchell Luskin, and R. Kent Smith. The technical program consisted of three-part tutorial lectures by Robert Dutton (Stanford), Alberto Sangiovanni-Vincentelli (Berkeley) and Mark Pinto (AT&T Bell Laboratories), followed by twelve invited one-hour lectures on the state of the art in VLSI simulation. The one-hour lectures were given by James Blue (National Bureau of Standards), Franco Brezzi (Pavia), Pierre Degond (Ecole Polytechnique), Joseph Jerome (Northwestern), Thomas Kerkhoven (Urbana), Peter Markowich (Vienna), Hans Mittelmann (Arizona State), Linda Petzold (Lawrence Livermore Laboratory), Conor Rafferty (AT&T Bell Laboratories), Christian Ringhofer (Arizona State), Donald Rose (Duke), and Thomas Seidman (University of Maryland, Baltimore). This volume represents contributions by the invited speakers, and in one instance, participants in the seminar.

Numerical simulation is rapidly becoming an important part of the VLSI design process, allowing the engineer to test, evaluate and optimize various aspects of chip design without resorting to the costly and time consuming process of fabricating prototypes. This of course accelerates the design cycle, at the same time improving the end result, since it is economically feasible to numerically simulate many more options than might otherwise be considered.

Numerical simulation tools fall into three general areas. First, there is *process simulation*, in which the physical processes required to fabricate a chip are modelled. The modelling here is usually done on the level of a single device, or a simple structure. The numerical model is a system of time-dependent, nonlinear partial differential equations, usually in two space dimensions.

The output of such a simulation serves as an input for *device simulation*, in which the electrical response of one or several devices on a chip are modelled under various operating conditions. The numerical model is a coupled set of nonlinear partial differential equations. These equations may be either elliptic, in the case of time-independent problems, or form a differential-algebraic system in the time-dependent case. As with process simulation, models in two space dimensions are most common. Output of the process model enters through the definition of the geometry of the devices, and the material properties, e.g., the doping profile.

With the computing power of the current generation of supercomputers, three space dimensional models are now beginning to be developed and used for both process and device simulation. Widespread use of such models will be an important step forward, since many of the physical phenomena being modelled are truly three dimensional in character.

The output of device simulation serves as input for *circuit simulation*. Unlike device and process simulation, circuit simulation attempts to model larger pieces of a chip, involving the interactions of many devices. The numerical model is a system of ordinary differential and algebraic equations. Normally, device simulations provide numerical approximations to nonlinear functions approximating terminal currents as a function of terminal voltages, which appear in this system. Of the three areas, circuit simulation is the most mature, both in terms of the sophistication of the models, and the state of software development.

The organizers would like to acknowledge and thank the IMA and the AMS for their administrative support and the Air Force Office of Scientific Research and the National Science Foundation for their financial support.

Lectures in Applied Mathematics
Volume **25** (1990)

Spatial-Temporal Structure of Solutions of the Semiconductor Device Problem

PETER A. MARKOWICH

Abstract. The semiconductor device equations are analysed by means of singular perturbation theory. It is shown that the solutions vary on two different time scales and on two different spatial scales (slow and fast ones) locally, i.e., thin temporal and spatial layers, within which the solutions vary rapidly, occur. Particular emphasis is given to the analysis of the structure of solutions within the temporal initial layer. In this layer the evolution process is dominated by electric field driven convection, and the governing equations become hyperbolic as diffusion approaches zero. The local hyperbolicity of the semiconductor device equations is used to show that spatial layers are generated on the fast time scale even when the initial data vary slowly.

1. Introduction. The semiconductor device equations, which model current flow, carrier transport, and potential distribution in (arbitrary) semiconductor devices, are given by the following system of p.d.e.'s:

(1.1)(a)	$\lambda^2 \Delta\psi = n - p - C(x)$	Poisson's equation
(1.1)(b)	$\dfrac{\partial n}{\partial t} = \operatorname{div} J_n + R$	electron continuity equation
(1.1)(c)	$\dfrac{\partial p}{\partial t} = -\operatorname{div} J_p + R$	hole continuity equation
(1.1)(d)	$J_n = \mu_n(\operatorname{grad} n - n \operatorname{grad}\psi)$	electron current relation
(1.1)(e)	$J_p = -\mu_p(\operatorname{grad} p + p \operatorname{grad}\psi)$	hole current relation.

1980 *Mathematics Subject Classification* (1985 *Revision*). Primary 35B25, 35B40, 35L60, 35P20, 78A35.

The device equations were derived by Van Roosbroeck [22] for the first time. The system (1.1) is already cast in scaled form (see Markowich [8] for details on the scaling), i.e., all quantities are dimensionless.

The scaled dependent variables are

ψ : electrical potential,
n : electron concentration,
p : hole concentration,
J_n : electron current density,
J_p : hole current density.

$C(x)$ denotes the scaled doping profile, which to a great extent determines the performance of the device under consideration. R is the scaled recombination-generation rate, and $\mu_n(\mu_p)$ the scaled electron (hole) mobility.

The semiconductor device is assumed to occupy the (bounded) domain $\Omega \subseteq \mathbf{R}^3$ ($\Omega \subseteq \mathbf{R}^2$ or even $\Omega \subseteq \mathbf{R}$ may be assumed for the simulation of certain devices). $x \in \Omega$ denotes the space variable and $t \geq 0$ the time variable, i.e., the dependent variables are considered as functions of x and t, e.g., $\psi = \psi(x,t)$.

Generally, $\partial\Omega$ splits into insulating segments $\partial\Omega_N$, on which the outward electric field component and the outward current density components vanish:

$$(1.1)(\mathrm{f}) \qquad -\operatorname{grad}\psi \cdot \nu = J_n \cdot \nu = J_p \cdot \nu = 0 \quad \text{on } \partial\Omega_N,\ t \geq 0$$

(ν denotes the exterior unit normal vector of $\partial\Omega$), and Ohmic contact segments $\partial\Omega_D$, on which Dirichlet data are prescribed:

$$(1.1)(\mathrm{g}) \qquad \psi = \psi_D,\ n = n_D,\ p = p_D \quad \text{on } \partial\Omega_D,\ t \geq 0.$$

Other types of boundary conditions are discussed in [15].

Initial data have to be specified for n and p, too:

$$(1.1)(\mathrm{h}) \qquad n(x,t=0) = n_I(x), \qquad p(x,t=0) = p_I(x), \qquad x \in \Omega.$$

Since n and p are physical concentrations, we require $n \geq 0$, $p \geq 0$, and, thus n_D, p_D, n_I, $p_I \geq 0$ has to hold.

The parameter λ, whose square appears as multiplier of the Laplace operator in Poisson's equation (1.1)(a) is the characteristic normed Debye length of the device. In most practical cases λ is very small, i.e., $\lambda \ll 1$. Since λ^2 is the coefficient of a derivative of largest occurring order, the problem (1.1) is singularly perturbed (see, e.g., Eckhaus [6], O'Malley [13]).

By inserting the current relations (1.1)(d), (e) into the corresponding continuity equations (1.1)(b), (c) and into the Neumann boundary conditions (1.1)(f), we obtain the following singularly perturbed elliptic-parabolic boundary value problem:

$$\left.\begin{aligned} &(1.2)(\mathrm{a}) \quad \lambda^2 \Delta\psi = n - p - C(x) \\ &(1.2)(\mathrm{b}) \quad n_t = \operatorname{div}(\mu_n(\nabla n - n\nabla\psi)) + R \\ &(1.2)(\mathrm{c}) \quad p_t = \operatorname{div}(\mu_p(\nabla p + p\nabla\psi)) + R \end{aligned}\right\} \quad x \in \Omega, \qquad t > 0,$$

$$(1.2)(\mathrm{d}) \qquad (\psi, n, p)|_{\partial\Omega_D} = (\psi_D, n_D, p_D)|_{\partial\Omega_D}, \qquad t > 0,$$

$$(1.2)(\mathrm{e}) \qquad \frac{\partial}{\partial\nu}(\psi, n, p)|_{\partial\Omega_N} = (0,0,0), \qquad t > 0,$$

$$(1.2)(\mathrm{f}) \qquad n(x, t=0) = n_I(x), \qquad p(x, t=0) = p_I(x), \qquad x \in \Omega.$$

In the last decade the semiconductor device problem has received quite a lot of attention from mathematicians, physicists, and device engineers. An excellent review of the underlying physics including an evaluation of various models for the physical parameters can be found in Selberherr [**15**]. For analytical results and for a discussion of numerical methods we refer the reader to the books by Mock [**12**] and Markowich [**8**].

In this paper we shall concentrate on the analysis of the structure of solutions based on "spatial" and "temporal" singular perturbations ($\lambda \to 0+$). Results on the solution structure are extremely important for the construction and analysis of efficient numerical methods (see Markowich [**8**]).

Steady state singular perturbation theory for the device problem is in a quite healthy state (see Brezzi [**1**]; Brezzi, Capelo, and Marini [**2**], Markowich and Ringhofer [**9**], Markowich [**7**], [**8**]). Only recently light has been shed on "temporal" asymptotics (see Ringhofer [**14**], Szmolyan [**20**], [**21**]). We shall present a state-of-the-art report, analyse in detail the multiple time scale variation of transient solutions, and give a new description of fast initial transients.

2. Spatial layers. Usually, the device domain Ω ($\subseteq \mathbf{R}^k$) splits into subdomains in which the doping profile $C(x)$ is positive (n-domains) and into subdomains in which $C(x)$ is negative (p-domains). The "inner" boundaries of these domains are called p-n junctions. Typically, the doping profile varies rapidly across the p-n junctions. For the sake of simplicity we shall make the "abrupt junction approximation," i.e., we shall assume that the doping profile has jump-discontinuities at the junctions (see Figure 2.1).

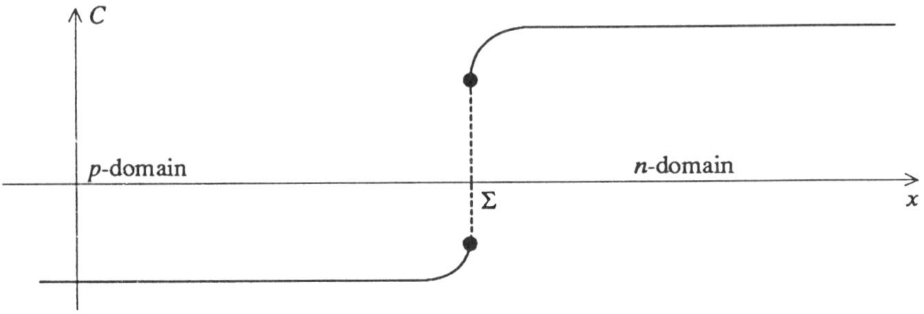

FIGURE 2.1. Abrupt junction approximation

The singular perturbation approach can also be applied to analyse devices with steeply graded junctions (see Markowich [**8**]). The grading determines the solutions locally at the junctions only.

Let the $(k-1)$-dimensional surface $\Sigma \subseteq \Omega$ be such an abrupt junction. Setting $\lambda = 0$ in (1.2)(a) gives the reduced equation

$$(2.1) \qquad 0 = n(\lambda = 0) - p(\lambda = 0) - C(x), \qquad x \in \Omega$$

(zero-space-charge approximation).

Since $[C]_\Sigma \neq 0$ ($[f]_\Sigma$ denotes the height of the jump of the function f at the surface Σ), we conclude that at least one of the functions $n(\lambda = 0)$, $p(\lambda = 0)$ has to be discontinuous at Σ. However, the solutions n and p of the "full" singularly perturbed problem (1.2) ($\lambda > 0$) are continuous in Ω because of the regularizing effect of elliptic and parabolic equations. Thus, as $\lambda \to 0+$, the reduced solutions $\psi(\lambda = 0)$, $n(\lambda = 0)$, $p(\lambda = 0)$ cannot be uniform approximations for the "full" solutions ψ, n, p.

This loss of regularity in going from the "full" Poisson's equation (1.2)(a) ($\lambda > 0$) to the reduced algebraic equation (2.1) by setting $\lambda = 0$ is typical for singularly perturbed problems.

A close mathematical scrutiny of the coupled steady state problem (by means of matched asymptotic expansions) shows that the potential ψ and the carrier concentrations n, p have thin (internal) layers at abrupt junctions within which they vary extremely rapidly (their gradients are $O(\frac{1}{\lambda})$ at junctions). The width of the layers is $O(\lambda |\ln \lambda|)$. Away from the junctions the solutions vary moderately and they are (at least in the steady state case) approximated to order λ by the reduced solutions (i.e., by the solutions of (1.2) with $\lambda = 0$ supplemented by appropriate interface conditions at the junctions).

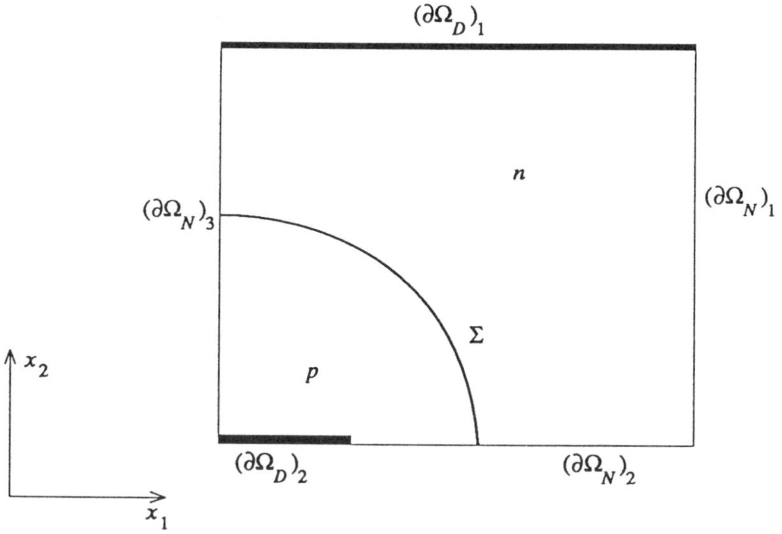

FIGURE 2.2. *p-n* diode simulation geometry

Boundary layers do not occur in our set-up since the Dirichlet data (1.2)(d) satisfy the zero-space-charge approximation:

$$0 = n_D - p_D - C(x), \qquad x \in \partial\Omega_D \tag{2.2}$$

(see Markowich [**8**] for the modelling of Ohmic contacts).

Figure 2.2 depicts the two-dimensional simulation geometry of a *p-n* diode and Figure 2.3 a typical steady state solution *p*. The rapid variation close to the *p-n* junction is apparent.

The internal spatial layers have a severe impact on the accuracy of numerical discretization methods. By now it is common knowledge in the area of numerical device simulation that "standard" finite difference and finite element discretizations of the continuity equations (1.2)(b),(c) give bad results on coarse grids (see Selberherr [**15**], Markowich [**8**]). Usually, exponentially fitted discretizations, which approximate the reduced problem well even on coarse meshes, are employed (Zlámal [**23**], [**24**], Markowich and Zlámal [**11**]).

An additional difficulty appears in the transient case. Obviously, the reduced problem (1.2) does not make sense at $t = 0$ unless the initial data n_I, p_I satisfy the zero-space-charge approximation (2.1). As we shall see in the next sections, this phenomenon leads to the occurrence of a thin initial temporal layer in the solutions ψ, n, p of the singularly perturbed semiconductor device problem.

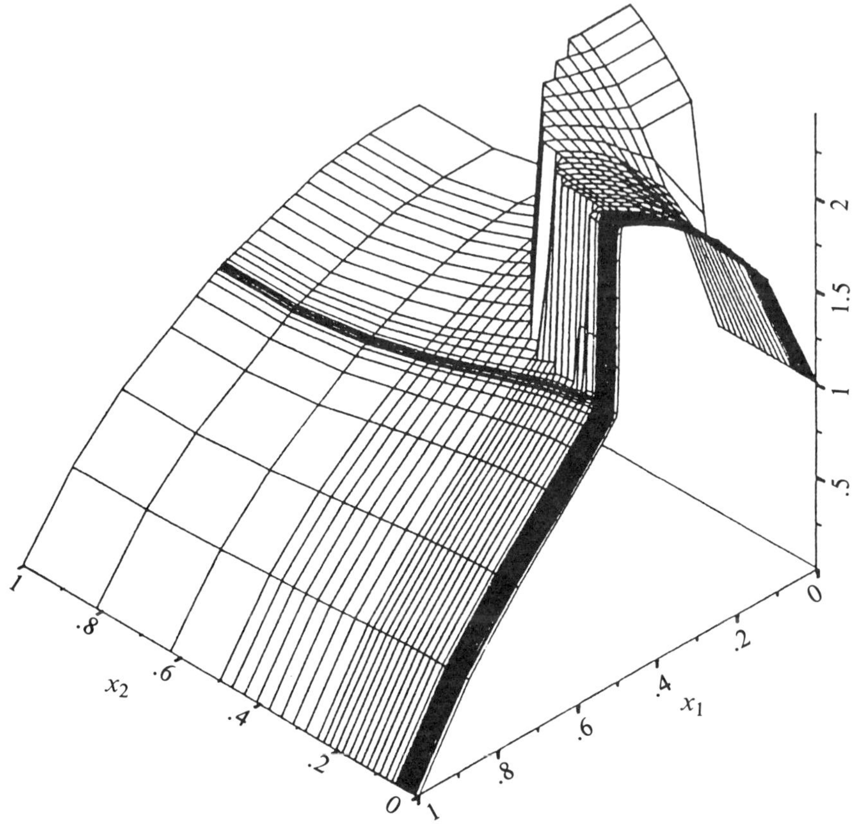

FIGURE 2.3. Hole concentration

3. Eigenvalues. In this section we shall motivate the occurrence of a fast time scale (on which transient solutions of (1.2) vary initially) by investigating the eigenvalues of the linearized problem (1.2).

A thorough analysis of the one-dimensional eigenvalue problem can be found in Steinrück [**17**], some results on the multidimensional case are given in Mock [**12**]. The linearized device problem was analysed (by means of semigroup theory) by Markowich and Ringhofer [**10**]. They obtained stability results for the differential problem and for the implicit Euler discretization.

The simplest semiconductor device is a resistor with homogeneous (scaled) doping $C(x) \equiv 1$, $x \in \Omega$. We assume for the sake of simplicity that the recombination-generation rate R vanishes identically and that $\mu_n \equiv \mu_p \equiv 1$. Then, for vanishing Dirichlet datum ψ_D, a steady state solution of (1.2) is given by

$$\psi^* \equiv 0, \qquad n^* \equiv n_D, \qquad p^* \equiv p_D \quad \text{in } \Omega \tag{3.1}$$

(for Ohmic contacts n_D and p_D only depend on the doping profile C at $\partial\Omega_D$, and therefore n_D, p_D are constant for the resistor).

We regard (1.2) as an evolution problem

$$\binom{n}{p}_t = F(n,p;\lambda), \qquad t > 0; \qquad \binom{n}{p}(t=0) = \binom{n_I}{p_I} \tag{3.2}$$

and linearize at (n^*, p^*) in direction (u, v):

$$\binom{u}{v}_t = DF(n^*,p^*;\lambda)\binom{u}{v}, \qquad t > 0, \tag{3.3)(a}$$

$$\binom{u}{v}(t=0) = \binom{u_I}{v_I}. \tag{3.3)(b}$$

For the sake of simplicity we shall only consider the homogeneous linearized problem.

The linearized operator $DF(n^*, p^*; \lambda)$ is given by

$$DF(n^*,p^*;\lambda)\binom{u}{v} = \binom{\Delta u - n_D\Delta\varphi}{\Delta v + p_D\Delta\varphi}, \qquad \lambda^2\Delta\varphi = u - v, \tag{3.3)(c}$$

$$(\varphi, u, v)|_{\partial\Omega_D} = \frac{\partial}{\partial\nu}(\varphi, u, v)|_{\partial\Omega_N} = 0. \tag{3.3)(d}$$

The corresponding eigenvalue problem

$$\sigma\binom{u}{v} = DF(n^*,p^*;\lambda)\binom{u}{v} \tag{3.4}$$

reads

$$\begin{aligned} &\text{(a)} \quad \sigma u = \Delta u - n_D\Delta\varphi, \\ &\text{(b)} \quad \sigma v = \Delta v + p_D\Delta\varphi, \\ &\text{(c)} \quad \lambda^2\Delta\varphi = u - v, \\ &\text{(d)} \quad (\varphi, u, v)|_{\partial\Omega_D} = \frac{\partial}{\partial\nu}(\varphi, u, v)|_{\partial\Omega_N} = 0. \end{aligned} \tag{3.5}$$

We multiply Poisson's equation (3.5)(c) by σ, use (3.5)(a), (b) and obtain

$$\sigma\lambda^2\Delta\varphi = \lambda^2\Delta^2\varphi - (n_D + p_D)\Delta\varphi. \tag{3.6}$$

We set $z := \Delta\varphi$ and derive the scalar eigenvalue problem:

$$\Delta z = \mu z, \quad x \in \Omega, \qquad \mu := \sigma + \frac{n_D + p_D}{\lambda^2}, \tag{3.7)(a}$$

(3.7)(b) $$z|_{\partial\Omega_D} = \frac{\partial z}{\partial \nu}|_{\partial\Omega_N} = 0.$$

Since the $(k-1)$-dimensional Lebesgue measure of $\partial\Omega_D$ is assumed to be positive, there exists a sequence of eigenvalues $\mu_i < 0$ of (3.7) (with finite multiplicities) and, thus, we obtain a sequence $\{\sigma_i^{(1)}\}_{i\in\mathbf{N}}$ of eigenvalues of (3.4) given by:

(3.8) $$\sigma_i^{(1)} = \mu_i - \frac{n_D + p_D}{\lambda^2}, \qquad i \in \mathbf{N}.$$

Another sequence of eigenvalues is obtained by setting $z = \Delta\varphi = 0$. Then (3.5) gives

(3.9) $$\sigma_i^{(2)} = \mu_i, \qquad i \in \mathbf{N}.$$

Clearly, the spectrum of $DF(n^*, p^*; \lambda)$ is the set $\{\sigma_i^{(1)}, \sigma_i^{(2)}\}_{i\in\mathbf{N}}$. Since DF^{-1}: $(L^2(\Omega))^2 \to (L^2(\Omega))^2$ is selfadjoint with respect to the scalar product

(3.10) $$(f, g)_D := \int_\Omega \left(\frac{f_1\overline{f_2}}{n_D} + \frac{g_1\overline{g_2}}{p_D}\right) dx, \qquad f = \begin{pmatrix} f_1 \\ f_2 \end{pmatrix}, \qquad g = \begin{pmatrix} g_1 \\ g_2 \end{pmatrix}$$

and compact, there exists a sequence

$$\left\{\begin{pmatrix} u_i^{(1)} \\ v_i^{(1)} \end{pmatrix}, \begin{pmatrix} u_i^{(2)} \\ v_i^{(2)} \end{pmatrix}\right\}_{i\in\mathbf{N}}$$

of orthogonal normed eigenfunctions. Thus, the solutions of the homogeneous linearized problem (3.3) are given by the Fourier series:

(3.11) $$\begin{pmatrix} u(t) \\ v(t) \end{pmatrix} = e^{-\frac{n_D+p_D}{\lambda^2}t} \sum_{i=1}^{\infty} \alpha_i \begin{pmatrix} u_i^{(1)} \\ v_i^{(1)} \end{pmatrix} e^{-\mu_i t} + \sum_{i=1}^{\infty} \beta_i \begin{pmatrix} u_i^{(2)} \\ v_i^{(2)} \end{pmatrix} e^{-\mu_i t}$$

with

$$\alpha_i = \left(\begin{pmatrix} u_i^{(1)} \\ v_i^{(1)} \end{pmatrix}, \begin{pmatrix} u_I \\ v_I \end{pmatrix}\right)_D, \qquad \beta_i = \left(\begin{pmatrix} u_i^{(2)} \\ v_i^{(2)} \end{pmatrix}, \begin{pmatrix} u_I \\ v_I \end{pmatrix}\right)_D.$$

For the simple device model (with steady state solution (3.1)) the eigenvalues μ_i and the eigenfunctions

$$\begin{pmatrix} u_i^{(2)} \\ v_i^{(2)} \end{pmatrix}$$

are independent of the singular perturbation parameter λ. In fact the eigenvalues (3.9) are obtained by setting $\Delta\varphi$ in (3.5) equal to zero, which corresponds to setting $\lambda = 0$. Thus, $\sigma_i^{(2)} = \mu_i$, $i \in \mathbf{N}$, are the eigenvalues of the linearized reduced problem.

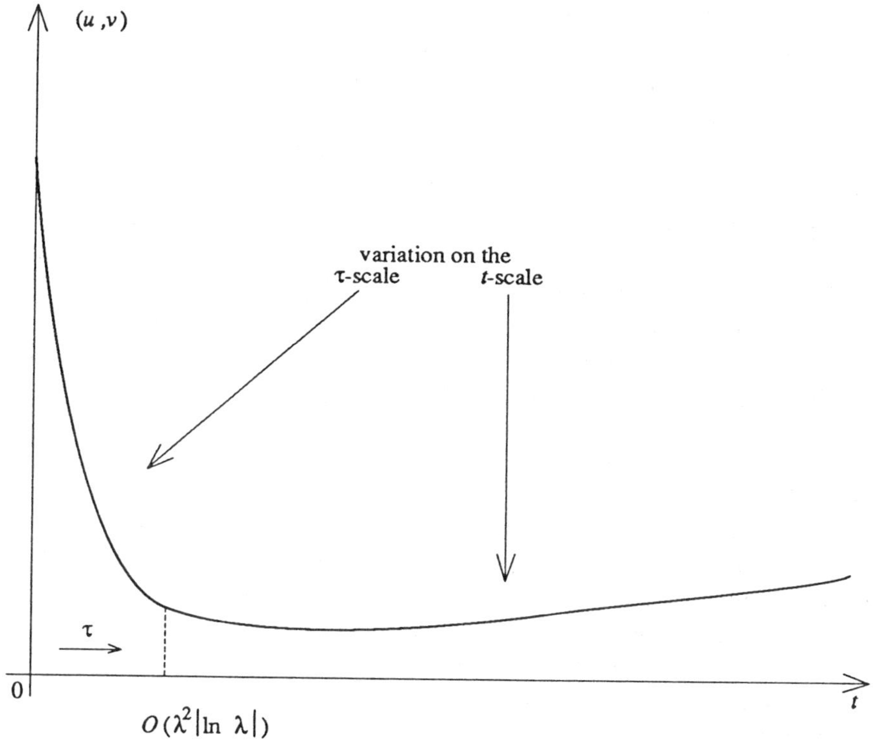

FIGURE 3.1. Initial layer

From (3.11) we conclude that the solution (u, v) of the initial value problem is of the following form:

$$\begin{pmatrix} u(t) \\ v(t) \end{pmatrix} = O(e^{-(n_D+p_D)t/\lambda^2}) + O(e^{-\mu_1 t}). \tag{3.12}$$

The first term on the right-hand side, which coresponds to the eigenvalues (3.8), decays on the "fast time scale"

$$\tau := \frac{t}{\lambda^2} \tag{3.13}$$

(the eigenfunctions, which correspond to low-index eigenvalues $\sigma_i^{(1)}$, are of low frequency, i.e., in particular the Fourier coefficient α_1 is $O(1)$ for slowly varying data!) and the second term decays on the "slow t-time scale." Therefore, the solutions of the linearized problem have initial-layer structure (see Figure 3.1). The width of the (temporal) initial layer is $O(\lambda^2 |\ln \lambda|)$.

The singular perturbation character of the device problem causes the "linearized" solutions of the resistor model to vary on two different time scales locally, namely on a fast one initially and, consecutively, on a slow one. The slow-time-scale solution corresponds to the reduced problem ($\lambda = 0$). Physically, the fast time scale is the dielectric relaxation time scale (see Sze **[19]**).

For linearizations of more complicated realistic device models (diodes, transistors, etc.) the eigenvalues cannot be computed explicitly. For the one-dimensional case it was shown by Steinrück **[17]** that the eigenvalues of the reduced linearized problem (which are responsible for the slow time scale) are approximations of eigenvalues of the "full" singularly perturbed problem if λ is small.

Because of the occurrence of negative $O(1/\lambda^2)$-eigenvalues (with corresponding low-frequency eigenfunctions), we conclude that the transient device problem is—in general—very stiff. Thus, we recommend L-stable time integrators (see Stetter **[18]**) in order to guarantee an accurate approximation of the slow-time-scale behaviour of solutions on coarse temporal grids (see Markowich and Ringhofer **[10]** for a stability analysis of the implicit Euler discretizations).

In the next sections we shall analyze the spatial and temporal structure of solutions of the (nonlinear) device problem within the thin temporal initial layer.

4. The hyperbolic initial layer problem. We insert the initial data n_I, p_I into Poisson's equation (1.2)(a) and obtain the following boundary value problem for the initial potential $\psi_I := \psi(t = 0)$:

$$\text{(4.1)(a)} \qquad \lambda^2 \Delta \psi_I = n_I - p_I - C(x), \qquad x \in \Omega,$$

$$\text{(4.1)(b)} \qquad \psi_I|_{\partial\Omega_D} = \psi_D(t = 0)|_{\partial\Omega_D}, \qquad \frac{\partial \psi_I}{\partial \nu}|_{\partial\Omega_N} = 0.$$

If n_I, p_I do not even approximately satisfy the zero-space-charge condition

$$\text{(4.2)} \qquad 0 = n_I - p_I - C(x), \qquad x \in \Omega,$$

then ψ_I is large in Ω, i.e.,

$$\text{(4.3)} \qquad \psi_I = O\left(\frac{1}{\lambda^2}\right).$$

This suggests rescaling the potential for $t \ll 1$. We set

$$\text{(4.4)} \qquad \psi = \frac{\varphi}{\lambda^2}$$

and expect the rescaled potential φ to be $O(1)$ initially.

Furthermore, we transform the device problem (1.2) "to the fast time scale"

$$\tau = \frac{t}{\lambda^2} \tag{4.5}$$

and obtain

$$\Delta\varphi = n - p - C(x), \tag{4.6)(a}$$

$$n_\tau = \lambda^2 \operatorname{div}(\mu_n \operatorname{grad} n) - \operatorname{div}(\mu_n n \operatorname{grad} \varphi) + \lambda^2 R, \tag{4.6)(b}$$

$$p_\tau = \lambda^2 \operatorname{div}(\mu_p \operatorname{grad} p) + \operatorname{div}(\mu_p p \operatorname{grad} \varphi) + \lambda^2 R, \tag{4.6)(c}$$

$$(\varphi, n, p)|_{\partial\Omega_D} = (\lambda^2 \psi_D, n_D, p_D)|_{\partial\Omega_D}, \quad \tau > 0, \tag{4.6)(d}$$

$$\frac{\partial}{\partial\nu}(\varphi, n, p)|_{\partial\Omega_N} = (0,0,0), \quad \tau > 0, \tag{4.6)(e}$$

$$n(x, \tau = 0) = n_I(x), \qquad p(x, \tau = 0) = p_I(x), \quad x \in \Omega. \tag{4.6)(f}$$

The singular perturbation parameter λ does not appear in Poisson's equation anymore; instead, λ^2 plays the role of a small diffusion parameter in the continuity equations (*on the fast-time scale*!).

Also, equations (4.6)(b), (c) show that the impact of recombination-generation of carriers on the fast-time-scale evolution of solutions is small, precisely $O(\lambda^2 \|R\|_{L^\infty(\Omega)})$.

Diffusion is small in the initial transient phase. By taking λ to the limit 0 (which corresponds to neglecting diffusion on the fast-time scale) we obtain the first-order convection equations:

$$\left.\begin{aligned} &(4.7)(a) \qquad n_\tau = -\operatorname{div}(\mu_n n \operatorname{grad} \varphi) \\ &(4.7)(b) \qquad p_\tau = \operatorname{div}(\mu_p p \operatorname{grad} \varphi) \end{aligned}\right\} \quad x \in \Omega, \qquad \tau > 0,$$

which are coupled to Poisson's equation (4.6)(a). Thus, in the limit $\lambda = 0$, the parabolic system (4.6) with small diffusion becomes hyperbolic ("zero" diffusion) on the fast-time scale. In this respect the transient device equations are in complete analogy to problems from various other areas of applied mathematics, e.g., fluid dynamics (Navier–Stokes equations for large Reynold number flow → inviscid Euler equations) or gas dynamics (see Smoller **[16]**).

By reasoning based on the vanishing viscosity method (see Smoller **[16]**), the hyperbolic system (4.7) coupled to Poisson's equation (4.6)(a) can be expected to describe the evolution of transient solutions within

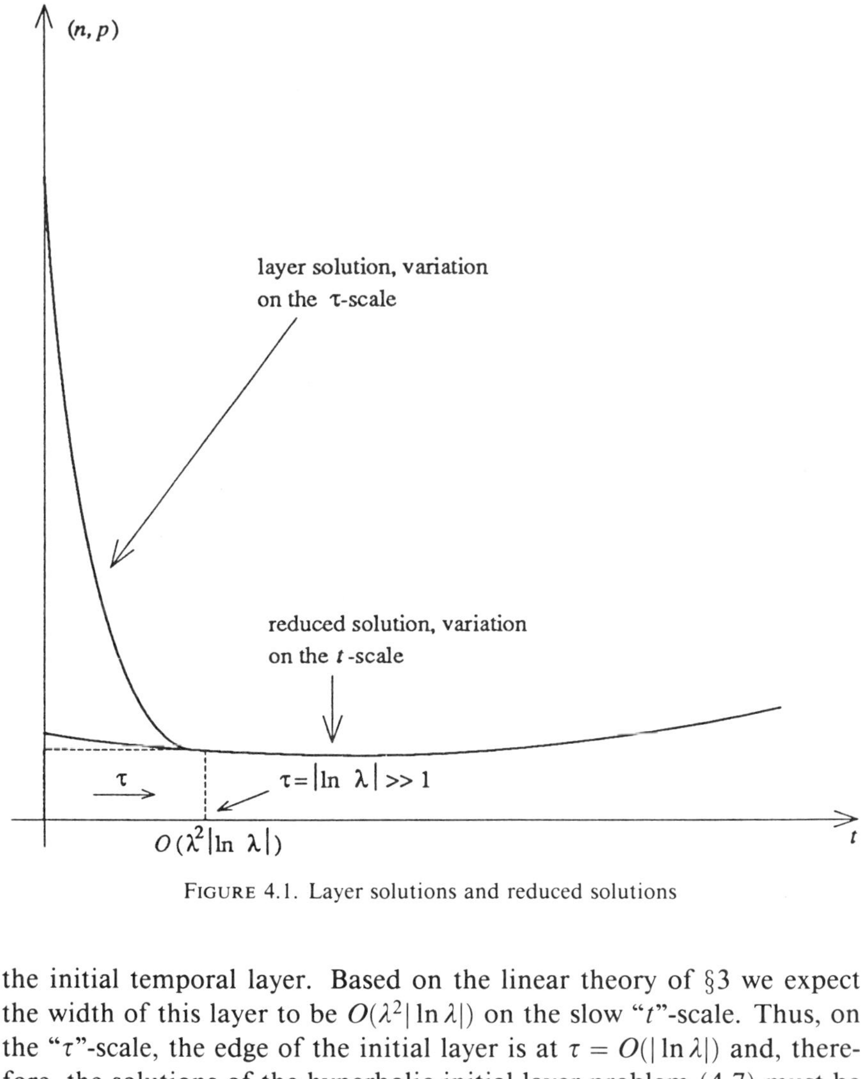

FIGURE 4.1. Layer solutions and reduced solutions

the initial temporal layer. Based on the linear theory of §3 we expect the width of this layer to be $O(\lambda^2|\ln\lambda|)$ on the slow "t"-scale. Thus, on the "τ"-scale, the edge of the initial layer is at $\tau = O(|\ln\lambda|)$ and, therefore, the solutions of the hyperbolic initial layer problem (4.7) must be available for *all* $\tau > 0$ in order to describe the initial transient on the "t"-scale for arbitrarily small values of λ. In particular, the steady state solutions of (4.7) ($\tau = \infty$!) can be expected to serve as good approximations of the initial data for the slow "t"-time scale evolution of n and p (see Figure 4.1), which is—at least outside spatial p-n layers—described by the reduced problem (1.2) ($\lambda = 0$ in Poisson's equation (1.2)(a)). Because of this the steady state solutions ($\tau = \infty$) of the hyperbolic layer problem (4.7) are of particular interest.

The problem (4.7), (4.6)(a) in the one-dimensional case was analyzed extensively by Szmolyan [**21**]. In this paper we shall concentrate on one particular issue, namely on the development of junction-discontinuities in n and p at $\tau = \infty$ from smooth initial data n_I, p_I. These discontinuities correspond to p-n junction layers which—as will be shown in the sequel—are generated within the thin temporal initial layer and then evolve on the slow "t"-time scale.

5. A model problem. As a simple device example we take a one-dimensional "symmetric" p-n diode, i.e., the device domain is the interval $(-1, 1)$ and the doping profile C is given by

$$(5.1)\qquad C(x) = \begin{cases} -1, & x \in (-1, 0) \quad (p\text{-domain}) \\ 1, & x \in (0, 1) \quad (n\text{-domain}) \end{cases}.$$

The p-n junction is at $x = 0$.

Furthermore, we assume $\psi_D(-1) = -\psi_D(1)$, $n_I(x) = p_I(-x)$, $n_D(-1) = p_D(1)$, $n_D(1) = p_D(-1)$, and $\mu_n(x) \equiv \mu_p(x) \equiv 1$. Then, by restraining to solutions which satisfy $\varphi(x) = -\varphi(-x)$, $p(-x) = n(x)$, the problem (4.7), (4.6)(a) is reduced to the n-domain $(0, 1)$:

$$\left.\begin{aligned} &(5.2)(\text{a}) \qquad \varphi_{xx} = n - p - C \\ &(5.2)(\text{b}) \qquad n_\tau = -(n\varphi_x)_x \\ &(5.2)(\text{c}) \qquad p_\tau = (p\varphi_x)_x \end{aligned}\right\} \qquad x \in (0, 1), \qquad \tau > 0$$

subject to the initial conditions

$$(5.2)(\text{d}) \quad n(x, \tau = 0) = n_I(x), \qquad p(x, \tau = 0) = p_I(x), \qquad x \in (0, 1)$$

and the Dirichlet boundary conditions

$$(5.2)(\text{e}) \qquad \varphi(0, \tau) = 0, \qquad \varphi(1, \tau) = \varphi_1, \qquad \tau > 0$$

where, by (4.6)(d), $\varphi_1 = \lambda^2 \psi_D(1)$. Thus, $\varphi_1 = 0$ holds in the limit $\lambda = 0$ unless the potential, which is externally applied at the contact $x = 1$, is of the order of magnitude λ^{-2}.

We simplify the problem (5.2) further by assuming

$$(5.3) \qquad p_I(x) = 0, \qquad x \in (0, 1).$$

By symmetry this implies that the initial datum n_I vanishes identically on the p-domain $(-1, 0)$.

The symmetry assumptions and (5.3) are illustrated in Figure 5.1.

From (5.3) and (5.2)(c) we conclude $p \equiv 0$ on $(0, 1)$ and then the initial layer problem (5.2) reads

$$\left.\begin{aligned} &(5.4)(\text{a}) \qquad \varphi_{xx} = n - 1 \\ &(5.4)(\text{b}) \qquad n_\tau = -(n\varphi_x)_x \end{aligned}\right\} \quad x \in (0, 1), \quad \tau > 0,$$

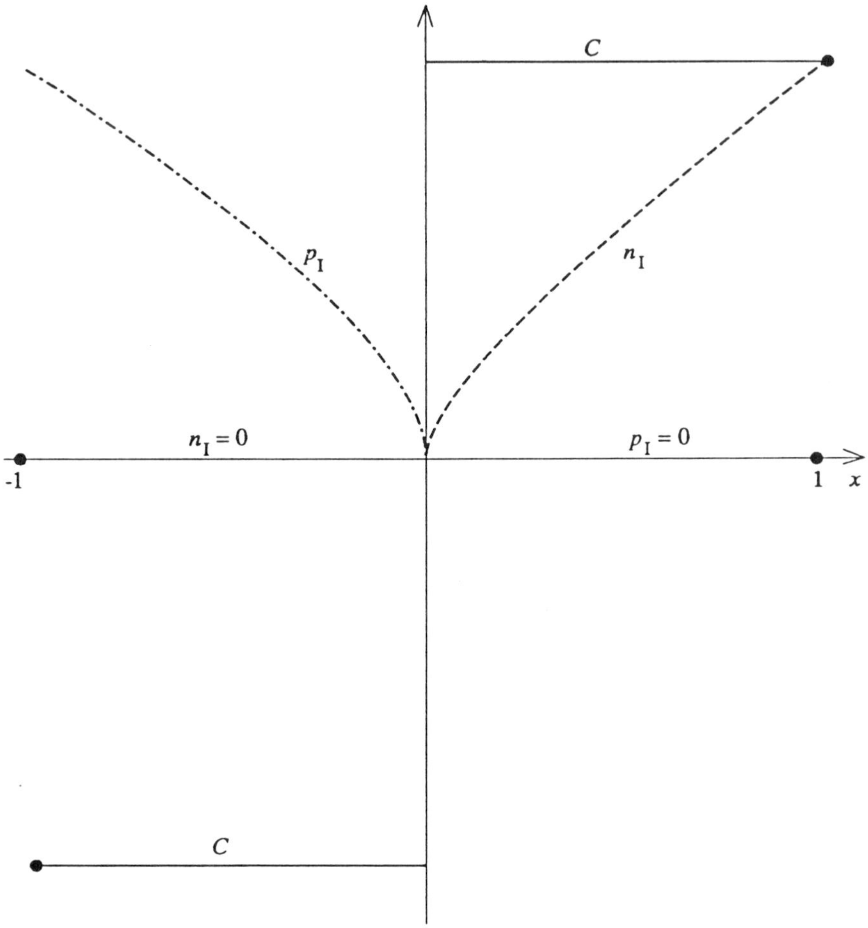

FIGURE 5.1. "Symmetric" diode

(5.4)(c) $$n(x, 0) = n_I(x), \quad x \in (0, 1),$$

(5.4)(d) $$\varphi(0, \tau) = 0, \quad \varphi(1, \tau) = \varphi_1 \in \mathbf{R}, \quad \tau > 0.$$

Firstly, we differentiate the right-hand side of (5.4)(b), use (5.4)(a) and obtain

(5.5) $$n_\tau + n_x \varphi_x = -n(n-1), \qquad x \in (0, 1), \qquad \tau > 0.$$

As usual in the analysis of hyperbolic problems we define the characteristic curve $\tilde{x}$ through the point $(x_0, \tau_0) \in [0, 1] \times \mathbf{R}_0^+$ as solution of the initial value problem (IVP):

(5.6)(a) $$\dot{\tilde{x}} = \varphi_x(\tilde{x}, \tau), \quad \tau > \tau_0,$$

(5.6)(b) $$\tilde{x}(\tau_0) = x_0$$

("dot" denotes differentiation with respect to τ). If the solution n satisfies

(5.7) $n(\cdot, \tau) \in L^\infty(0,1)$ uniformly on bounded τ-intervals

then (5.4)(a) implies that the IVP (5.6) is uniquely solvable ($\varphi_x(\cdot, \tau) \in W^{1,\infty}(0,1)$ uniformly on bounded τ-intervals!). Thus, different characteristics do not intersect.

The equation (5.5) can now be written in characteristic form

(5.8) $$\frac{d}{d\tau}\tilde{n} = -\tilde{n}(\tilde{n} - 1), \qquad \tau > \tau_0$$

where $\tilde{n} := n(\tilde{x}(\tau), \tau) \cdot \frac{d}{d\tau}$ denotes the directional derivative along the characteristic $\tilde{x}$.

We study a Riemann-type problem, i.e., we take a discontinuous initial datum n_I:

(5.9) $$n_I(x) = \begin{cases} 0, & 0 \le x < X \\ 1, & X < x \le 1 \end{cases}, \qquad X \in (0,1).$$

Since $n = 0$, $n = 1$ are steady states of (5.8), the problem (5.4), (5.9) is solved when the curve $\Gamma : x = x_s(\tau)$ in the (x, τ)-plane, which "separates" the states $n = 0$ and $n = 1$, is determined.

The Rankine–Hugeniot condition (see Smoller **[16]**) for (5.4)(b) reads

(5.10) $$\dot{x}_s[n]_\Gamma = [n\varphi_x]_\Gamma.$$

Since φ_x is continuous across Γ (by (5.4)(a)) we conclude $[n\varphi_x]_\Gamma = [n]_\Gamma \varphi_x|_\Gamma$ and

(5.11) $$\dot{x}_s = \varphi_x|_\Gamma$$

follows. The separation curve Γ is the characteristic through $(X, 0)$:

(5.12)(a) $$\dot{x}_s = \varphi_x(x_s, \tau), \qquad \tau > 0,$$

(5.12)(b) $$x_s(0) = X.$$

The curve Γ splits the domain $(0,1) \times \mathbf{R}^+$ into two subdomains Λ_0 and Λ_1 (see Figure 5.2). $n \equiv 1$ in Λ_1 and $n \equiv 0$ in Λ_0 hold.

Poisson's equation (5.4)(a) reads

(5.13) $$\varphi_{xx} = \begin{cases} -1 \text{ in } \Lambda_0, & \varphi(0,\tau) = 0, \\ 0 \quad \text{in } \Lambda_1, & \varphi(1,\tau) = \varphi_1. \end{cases}$$

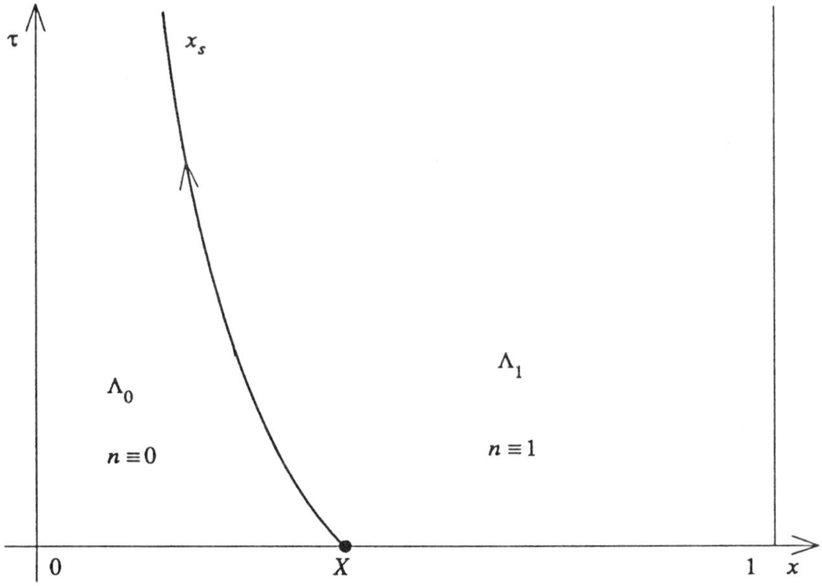

FIGURE 5.2. The Riemann problem

Straightforward integration and the continuity of φ and φ_x across Γ give

$$\varphi_x(x,\tau) = \begin{cases} -x + x_s(\tau) - \dfrac{x_s^2(\tau)}{2} + \varphi_1 & \text{in } \Lambda_0, \\ -\dfrac{x_s^2(\tau)}{2} + \varphi_1 & \text{in } \Lambda_1. \end{cases} \tag{5.14}$$

Then, the IVP (5.12), which determines x_s, reads

$$\dot{x}_s = -\frac{x_s^2}{2} + \varphi_1, \qquad \tau > 0, \tag{5.15)(a}$$

$$x_s(0) = X. \tag{5.15)(b}$$

By integrating (5.15) it becomes apparent that three cases must be distinguished:

(i) $\varphi_1 = 0 \quad \Rightarrow x_s(\tau) \sim \dfrac{2}{\tau}$ as $\tau \to \infty$,

(ii) $\varphi_1 > 0 \quad \Rightarrow x_s(\tau) = \sqrt{2\varphi_1} + O(e^{-\sqrt{2\varphi_1}\tau})$ as $\tau \to \infty$,

(iii) $\varphi_1 < 0 \quad \Rightarrow \exists \tau^* < \infty : 0 \le x_s(\tau) < 1$ for

$$0 \le \tau < \tau^* \text{ and } x_s(\tau^*) = 0$$

(see Figure 5.3). $\varphi_1 = 0$ corresponds to a moderate externally applied

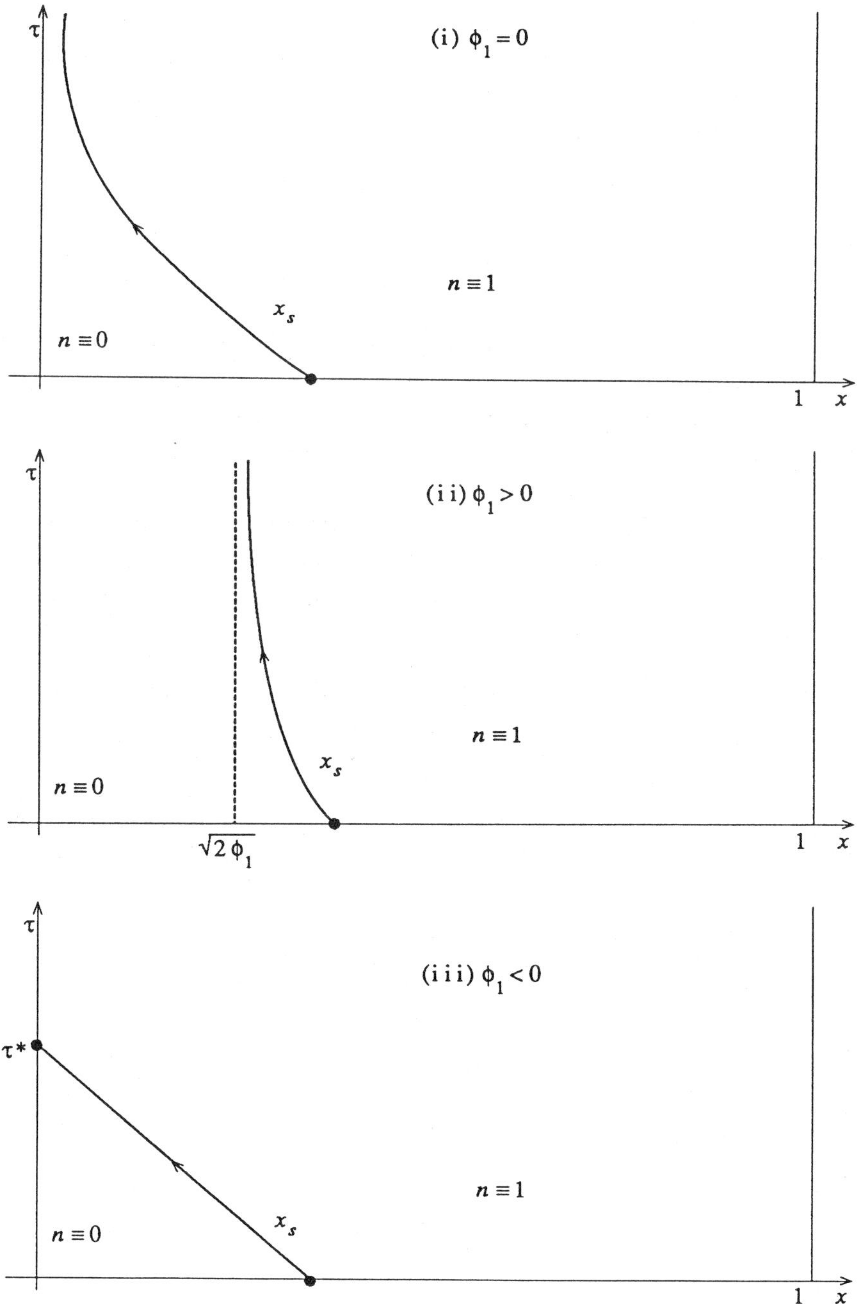

FIGURE 5.3. Solutions of the Riemann problem

potential in the limit $\lambda \to 0$. In this case the discontinuity propagates towards the junction and the steady state solution (on the τ-scale) is $n \equiv 1$, $x \in [0,1]$ (i.e., $n \equiv 0$ on $[-1,0)$ and $n \equiv 1$ on $(0,1]$ by symmetry).

$\varphi_1 > 0$ corresponds to a large externally applied reverse bias. The steady state solution then is

$$n(x, \tau = \infty) = \begin{cases} 0, & 0 \le x < \sqrt{2\varphi_1} \\ 1, & \sqrt{2\varphi_1} < x \le 1 \end{cases}, \qquad \varphi_1 > 0. \tag{5.16}$$

The interval $[0, \sqrt{2\varphi_1})$, and consequently $(-\sqrt{2\varphi_1}, \sqrt{2\varphi_1})$ by symmetry, is totally depleted of carriers (depletion region).

In the case of a large externally applied forward bias $\varphi_1 < 0$, the junction becomes flooded with carriers in finite (fast) time. $n \equiv 1$ for $\tau > \tau^*$ holds on $[0,1]$.

The results on the steady state solution obtained here for the Riemann problem were extended to general initial data. We only state the generalization of (i), (ii), a proof (based on the method of characteristics) can be found in Szmolyan [**21**].

THEOREM 5.1. *Let $n_I : (0,1] \to \mathbf{R}^+$, $n_I(0) = 0$, $n_I(1) = 1$, $n_I \in L^\infty(0,1)$, be such that the function φ_I defined by $\varphi_I'' = n_I - 1$, $\varphi_1(0) = 0$, $\varphi_I(1) = \varphi_1 \ge 0$ satisfies $\varphi_I'(0) > 0$, $\varphi_I'(1) < 0$. Also assume that $\varphi_1 \le \frac{(\varphi_I'(0) - \varphi_I'(1))^2}{2}$ holds. Then the problem* (5.4), (5.2)(e) *subject to the boundary conditions*

$$n(0,\tau) = 0, \qquad n(1,\tau) = 1, \qquad \tau > 0 \tag{5.17}$$

has a unique solution (φ, n) defined for $\tau \in [0, \infty)$, which satisfies

$$\lim_{\tau \to \infty} n(x,\tau) = \begin{cases} 0, & 0 \le x < \sqrt{2\varphi_1} \\ 1, & \sqrt{2\varphi_1} < x \le 1 \end{cases} \quad \text{a.e.} \tag{5.18}$$

The conditions $\varphi_I'(0) > 0$, $\varphi_I'(1) < 0$ imply that the characteristics starting at the boundaries $x = 0$ and $x = 1$ point inward initially.

Thus, boundary conditions for n must be prescribed initially. In the course of the proof of Theorem 5.1 it is shown that $\varphi_x(0,\tau) > 0$ and $\varphi_x(1,\tau) < 0$ hold for all $\tau \geq 0$, i.e., the global boundary conditions (5.17) are required.

It was shown by Brezzi, Capelo, and Marini [**2**] that the solutions of the stationary problem with small diffusion

$$\left.\begin{aligned} &\varphi_{xx} = n - 1 && (5.19)(\mathrm{a}) \\ &\lambda^2 n_{xx} - (n\varphi_x)_x = 0 && (5.19)(\mathrm{b}) \end{aligned}\right\} \quad x \in (0,1),$$

$$\varphi(0) = 0, \quad \varphi(1) = \varphi_1, \quad n(0) = 0, \quad n(1) = 1, \tag{5.19)(c}$$

converge to the steady state solution (5.18) of (5.4) as $\lambda \to 0+$ (if $\varphi_1 \leq 1/2$), i.e., the steady state solution on the τ-scale of the transient problem without diffusion is the limiting solution of the stationary unipolar device model as diffusion approaches zero. The relationship between the limits "$\lambda \to 0+$" and "$\tau \to \infty$" has not been analysed yet for bipolar models of the type (4.6) (so far, further results on the limit "$\lambda \to 0+$" for the stationary problem (4.6) were obtained by Brezzi [**1**]; Brezzi, Capelo, and Gastaldi [**3**] in the one-dimensional case; and by Caffarelli and Friedman [**4**] for multidimensional p-n diode models).

The bipolar hyperbolic system (5.2) was analyzed (for arbitrary doping profile C) by Szmolyan [**21**]. An existence and uniqueness result as well as results on the asymptotic behaviour as $\tau \to \infty$ were obtained.

In order to illustrate the main properties of the solutions in the bipolar case we present plots of numerically obtained solutions of (5.2) for a p-n diode with

$$C(x) = \begin{cases} -1, & x \in [0, \frac{1}{2}), \quad p\text{-region}, \\ 1, & x \in (\frac{1}{2}, 1], \quad n\text{-region}, \end{cases}$$

and $\varphi_1 = 0$ (moderate externally applied potential and $\lambda \to 0+$). The initial data n_I, p_I are smooth, moderately varying functions without layer structure. The corresponding (fast-time-scale) solutions $n = n(x,\tau)$, $p = p(x,\tau)$ are depicted in Figures 5.4 and 5.5, respectively. It is apparent that the variation of n and p locally at the junction $x = 1/2$

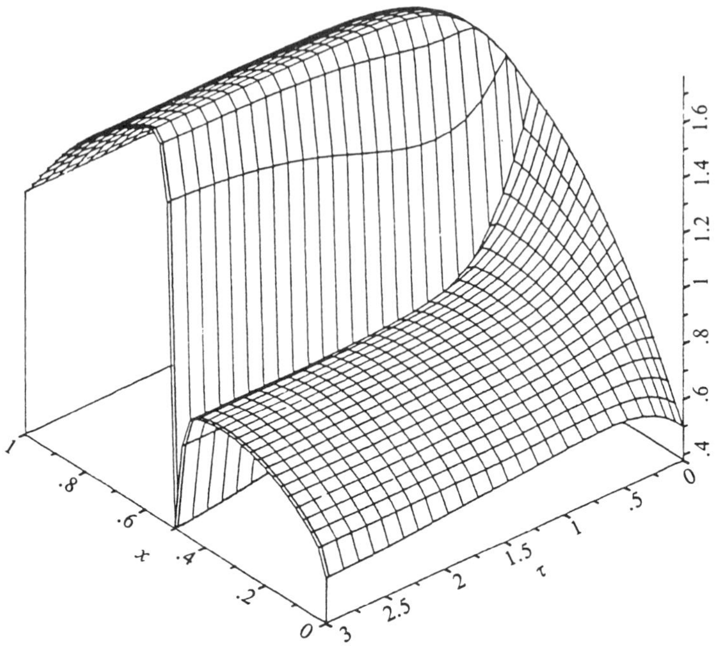

FIGURE 5.4. Electron concentration

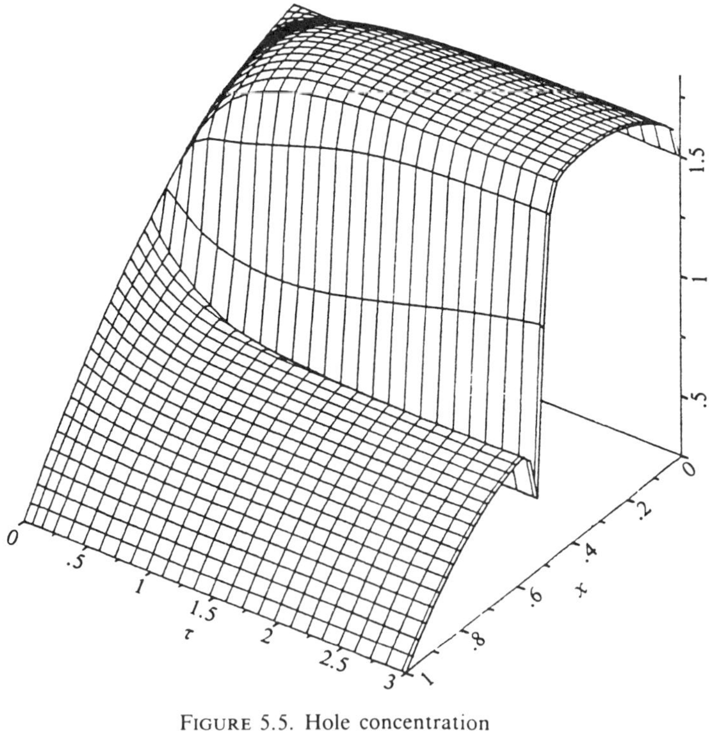

FIGURE 5.5. Hole concentration

becomes large after a few units of the τ-scale (the steady state solutions $n(x, \tau = \infty)$, $p(x, \tau = \infty)$ have jump-discontinuities at the junction). Thus, the "correct" spatial layer structure, which is a main feature of the slow-time-scale evolution and of steady state solutions ($t = \infty$), develops extremely fast (i.e., within the thin temporal initial layer of width $O(\lambda^2 |\ln \lambda|)$) even for "unstructured" initial data.

The (negative rescaled) field φ_x and the (negative) space charge $n - p - C$ are shown in Figures 5.6 and 5.7, respectively. Both decay to zero on the fast-time scale (dielectric relaxation time scale).

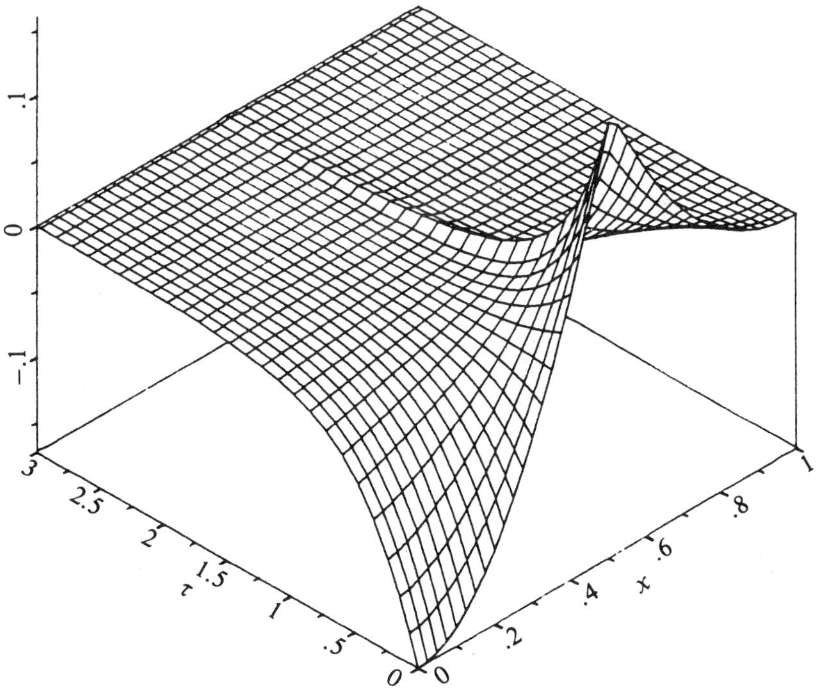

FIGURE 5.6. Electric field

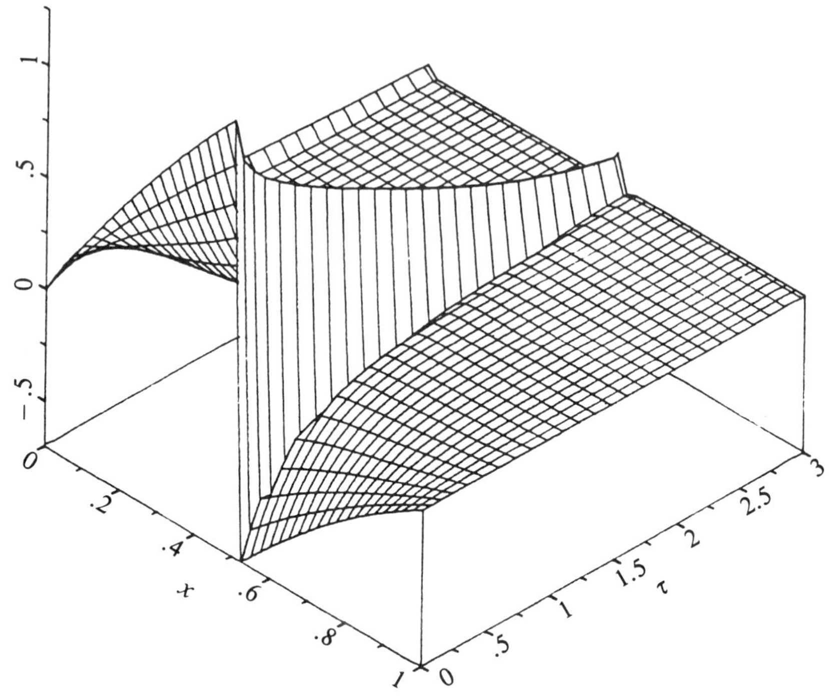

FIGURE 5.7. Space charge

6. Spatial-temporal solution structure and conclusions. The singular perturbation analysis of the semiconductor device equations leads to a qualitative and quantitative description of the transient evolution of solutions. The main features are as follows.

(a) Thin internal spatial layers within which the solutions ψ, n, p vary rapidly (as functions of the space variable) occur at abrupt or steeply graded p-n junctions. At an abrupt junction grad ψ, grad n, grad p are $O(\lambda^{-1})$ and the width of the layer is $O(\lambda|\ln\lambda|)$. The width of the layer at a graded junction depends on the grading, too. We remark that boundary layers with the analogous solution structure occur at semiconductor-oxide interfaces and at Schottky contacts (see Markowich **[8]**).

(b) Generally, a thin temporal initial layer of width $O(\lambda^2|\ln\lambda|)$ occurs at $t = 0$. Within this layer the solutions ψ, n, p vary fast as functions of time, i.e., ψ_t, n_t, p_t are maximally $O(\lambda^{-2})$. The spatial layers are generated (even for slowly varying initial data n_I, p_I) within the temporal layer.

The spatial-temporal solution structure is illustrated in Figure 6.1 for a one-dimensional device with one p-n junction at $x = 0$.

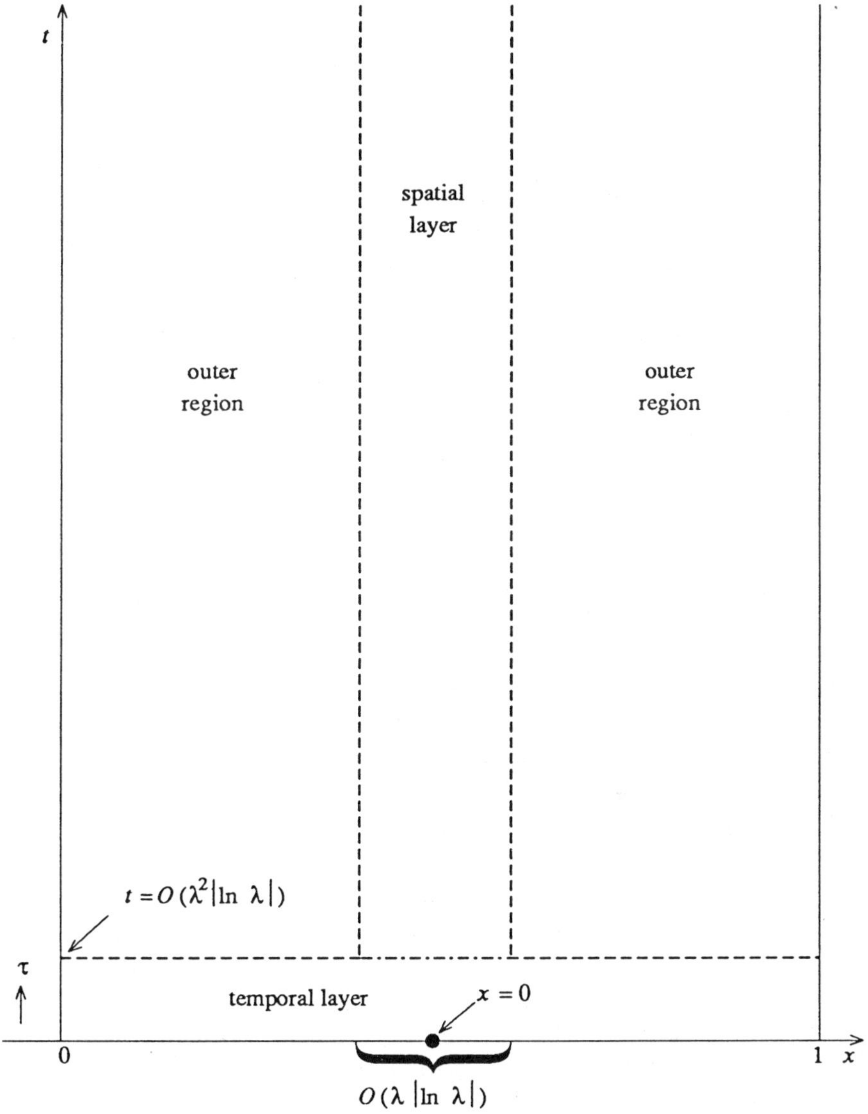

FIGURE 6.1. Solution structure

Within the temporal initial layer the solutions vary on the $\tau = t/\lambda^2$ scale and they are approximately obtained by solving the hyperbolic problem (4.7), (4.6)(a), (f) (subject to appropriate boundary conditions). After the fast-time-scale evolution the spatial p-n layer at $x = 0$

is formed and the evolution on the t-scale takes over. Within the p-n layer the solutions vary on the $\rho = \frac{x}{\lambda}$-scale and they are described by a system of ordinary differential equations with ρ as independent variable in which t only enters as a parameter (internal layer equations, see Markowich [7], Ringhofer [14]). In the two outer regions outside the spatial and temporal layers the solutions ψ, n, p vary slowly as functions of time and space. There, they are approximated by the solutions of the reduced equations ((1.2) with $\lambda = 0$). Interface conditions for the reduced equations are obtained by matching with the internal layer equations (see Markowich [8]) and the initial data are the steady state solutions ($\tau = \infty$) of the hyperbolic initial layer problem.

Often, stationary solutions, which already have the qualitatively correct internal layer structure, are taken as initial data n_I, p_I for transient simulations. Then $\psi_I = O(1)$ in Ω and, consequently, the rescaled potential φ defined in (4.4) is $O(\lambda^2)$. Also, $n_I - p_I - C = O(\lambda^2)$ holds outside p-n layers. Thus, in first approximation, the fast-time-scale solutions satisfy $n(x, \tau) = n_I(x)$, $p(x, \tau) = p_I(x)$ outside the spatial layers ($n = n_I, p = p_I, \varphi = 0$ are steady states of (4.7), (4.6)(a) if $n_I - p_I - C = 0$ holds). Therefore, the initial temporal layer does not appear explicitly in the solutions. A detailed analysis of this case (admitting small perturbations only) was given by Ringhofer [14], Szmolyan [20].

We remark that, even in the case when a temporal initial layer does not appear, the stiffness of the transient device problem is retained (see Section 3).

The complicated structure of transient solutions requires particular attention when a numerical discretization algorithm is designed. Exponentially fitted spatial discretizations, which are capable of resolving the solutions accurately on relatively coarse meshes are commonly used (see Markowich [8], Markowich and Zlámal [11]. L-stable time integration algorithms (see Stetter [18]), which cope with the "transient" stiffness in the sense that the correct slow-time-scale evolution is resolved well without an exceedingly fine grid within the initial layer, have proven to be most efficient (see Markowich and Ringhofer [10] for a stability analysis and Zlámal [24] for a spatial-temporal convergence analysis).

In a recent paper Douglas, Gamba, and Squeff [5] propose a discretization algorithm which is based on convection dominating diffusion. Methods of this type, which emphasize the local hyperbolicity, are particularly well suited to numerically resolving the fast-time-scale evolution. However, on the slow-time scale, diffusion generally dominates convection (particularly in outer regions), and, thus, "parabolic" discretization methods (of L-stable type) perform well.

Bibliography

1. F. Brezzi, *Theoretical and Numerical Problems in Reverse Biased Semiconductor Devices*, Proc. of the Seventh International Symposium on 'Computing Methods in Applied Sciences and Engineering', Versailles, France, Dec. 1985.

2. F. Brezzi, A. Capelo, and L. D. Marini, *Singular Perturbation Problems in Semiconductor Devices*, Proc. of the Meeting in Guanajuato, July 1985.

3. F. Brezzi, A. Capelo, and L. Gastaldi, *A Singular Perturbation Analysis for Semiconductor Devices*, submitted.

4. L. Caffarelli and A. Friedman, *Singular Perturbation Problem for Semiconductors*, to appear in Boll. U.M.I.

5. J. Douglas, Jr., I. M. Gamba, and M. C. J. Squeff, *Simulation of the Transient Behaviour of a One-Dimensional Semiconductor Device*, submitted.

6. W. Eckhaus, *Asymptotic Analysis of Singular Perturbations*, North Holland, Amsterdan-New York-Oxford, 1979.

7. P. A. Markowich, *A singular perturbation analysis of the fundamental semiconductor device equations*, SIAM J. Appl. Math. **44**, No. 5 (1984), 896–928.

8. P. A. Markowich, *The Stationary Semiconductor Device Equations*, Springer, Wien-New York, 1986.

9. P. A. Markowich and C. A. Ringhofer, *A singularly perturbed boundary value problem modelling a semiconductor device*, SIAM J. Appl. Math. **44**, No. 2 (1984), 231–256.

10. P. A. Markowich and C. A. Ringhofer, *Stability of the linearized transient semiconductor device equations*, ZAMM **67** (1987), 319–332.

11. P. A. Markowich and M. Zlámal, *Inverse-Average-Type Finite Element Discretisations of Self-Adjoint Second Order Elliptic Problems*, Math. Comp. **51** (1988), 431–449.

12. M. S. Mock, *Analysis of Mathematical Models of Semiconductor Devices*, Boole Press, Dublin, 1983.

13. R. E. O'Malley, Jr., *Introduction to Singular Perturbations*, Academic Press, New York, 1974.

14. C. A. Ringhofer, *An asymptotic analysis of a transient pn-junction model*, to appear in SIAM J. Appl. Math..

15. S. Selberherr, *Analysis and Simulation of Semiconductor Devices*, Springer, Wien-New York, 1984.

16. J. Smoller, *Shock Waves and Reaction-Diffusion Equations*, Springer, New York-Heidelberg-Berlin, 1983.

17. H. Steinrück, *Singularly perturbed eigenvalue problems*, to appear in SIAM J. Appl. Math.

18. H. J. Stetter, *Analysis of Discretisation Methods for Ordinary Differential Equations*, Springer, New York-Heidelberg-Berlin, 1973.

19. S. M. Sze, *Physics of Semiconductor Devices*, 2nd ed., J. Wiley, New York, 1981.

20. P. E. Szmolyan, *A Singular Perturbation Analysis of the Transient Semiconductor Device Equations*, submitted.

21. P. E. Szmolyan, *Ein hyperbolisches System aus der Halbleiterbauelementphysik*, Dissertation, TU-Wien, 1987.

22. W. V. Van Roosbroeck, *Theory of flow of electrons and holes in germanium and other semiconductors*, Bell. Syst. Tech. J. **29** (1950), 560–607.

23. M. Zlámal, *Finite element solution of the fundamental equations of semiconductor devices* I, Math. Com. **46**, No. 173 (1986), 27–43.

24. M. Zlámal, *Finite Element Solution of the Fundamental Equations of Semiconductor Devices* II, submitted.

Fachbereich Mathematik, TU-Berlin, Strasse des 17.juni 135, D-1000 Berlin 12, Federal Republic of Germany

Lectures in Applied Mathematics
Volume **25** (1990)

The Shape of Solutions to the Fundamental Semiconductor Device Equations

CHRISTIAN RINGHOFER

Abstract. The structure of solutions to the equations, arising from the two-dimensional model of an MOS translator, is analyzed via a singular perturbation approach. Inversion layer effects are discussed and different time scales are separated. The validity of the expansion is demonstrated with numerical examples.

1. Introduction. We give an overview of the treatment of the fundamental semiconductor equations via singular perturbation analysis and its implications on time differencing methods. The equations as first given in **[11]** are of the form

$$\text{(1.1)} \qquad \begin{aligned} &\text{(a)} \quad -\varepsilon\Delta\psi + q[n - p - C] = 0, \\ &\text{(b)} \quad -\operatorname{div}(\mathscr{J}_n) + q(\partial_t n + R) = 0, \\ &\text{(c)} \quad \operatorname{div}(\mathscr{J}_p) + q(\partial_t p + R) = 0, \\ &\text{(d)} \quad \mathscr{J}_n = q[D_n \nabla n - \mu_n n \nabla \psi], \\ &\text{(e)} \quad \mathscr{J}_p = q[-D_p \nabla p - \mu_p p \nabla \psi]. \end{aligned}$$

The dependent variables have the following meaning and units:

$$\text{(1.2)} \qquad \begin{aligned} \psi&: \quad \text{electric potential (V)} \\ n&: \quad \text{electron density (cm}^{-3}\text{)} \\ p&: \quad \text{hole density (cm}^{-3}\text{)} \\ \mathscr{J}_n&: \quad \text{electron current density (A/cm}^2\text{)} \\ \mathscr{J}_p&: \quad \text{hole current density (A/cm}^2\text{)}. \end{aligned}$$

1980 *Mathematics Subject Classification* (1985 *Revision*). Primary 35K55, 35K65.
Supported by NSF Grant # PMS 8801153.

The parameters have the following meaning and values for silicon at room temperature:

(1.3)
$$\begin{aligned} \varepsilon &: \quad \text{permitivity, } 10^{-12} \text{ (As/V)},\\ q &: \quad \text{unit charge, } 10^{-19} \text{ (As)}. \end{aligned}$$

$C(x)$ in (1.1)(a) is the doping concentration. R is the recombination rate. Various models can be used for R (see **[10]** for an overview).

For this paper we will consider the Shockley–Read–Hall recombination rate given by

(1.4)
$$R = (np - n_i^2)[\tau_n(n + n_i) + \tau_p(p + n_i)]^{-1}$$

where n_i is the intrinsic density ($\sim 10^{10}$). τ_n and τ_p are the average lifetimes for electrons and holes ($\sim 10^{-6}$ sec). D_n, D_p, μ_n, and μ_p denote the diffusion coefficients and mobilities for electrons and holes. They are either given as functions of space or as functions of the spatial variables and the electric field $(-\nabla\psi)$. Again we refer to **[10]** for an overview over different models. (1.1) will be posed on a domain Ω. The boundary $\partial\Omega$ of the domain will generally be decomposed into

(1.5)
$$\partial\Omega = \Gamma_D\ \Gamma_N\ \Gamma_G^1.$$

On Γ_D, which models Ohmic contacts, Dirichlet boundary conditions will be imposed. They are of the form

(1.6)
$$\begin{aligned} &\text{(a)} \quad \psi = V + \psi_{\text{bi}},\\ &\text{(b)} \quad n = n_b = \tfrac{1}{2}(C + \sqrt{C^2 + 4n_i^2}),\\ &\text{(c)} \quad p = p_b = \tfrac{1}{2}(-C + \sqrt{C^2 + 4n_i^2}), \qquad x \in \Gamma_D. \end{aligned}$$

V is the applied potential. ψ_{bi}, the built-in potential, is given by

(1.7)
$$\psi_{\text{bi}} = \ln[(2n_i)^{-1}(C + \sqrt{C^2 + 4n_i^2})].$$

Note that $n_b - p_b - C = 0$ and $n_b p_b = n_i^2$ holds. This means that at Ohmic contacts there is no charge and the carriers are in thermal equilibrium. At Γ_N, which models insulating or artificial surfaces, Neumann boundary conditions hold:

(1.8)
$$\nabla\psi \cdot \mathbf{n} = \nabla n \cdot \mathbf{n} = \nabla p \cdot \mathbf{n} = 0, \qquad \mathbf{x} \in \Gamma_N.$$

$\mathbf{n}$ denotes the normal vector on Γ_N. For some devices, like MOS translators, an additional region Ω^{Ox}, modeling a layer of oxide, is attached to Ω. In Ω^{Ox}

(1.9)
$$\Delta\psi^{\text{Ox}} = 0$$

holds. Γ_G^1 denotes the interface between Ω and Ω^{Ox}. There

$$\text{(1.10)} \qquad \begin{aligned} &\text{(a)} \quad \psi = \psi^{\mathrm{Ox}}, \\ &\text{(b)} \quad \nabla\psi \cdot \mathbf{n} = \omega \nabla \psi^{\mathrm{Ox}} \cdot \mathbf{n}, \\ &\text{(c)} \quad \mathscr{J}_n \cdot \mathbf{n} = \mathscr{J}_p \cdot \mathbf{n} = 0 \end{aligned}$$

hold. The rest of the boundary of Ω^{Ox} consists of a Neumann part Γ_N^{Ox} and a part Γ_G^2 where Dirichlet boundary conditions for ψ^{Ox} hold:

$$\text{(1.11)} \qquad \psi^{\mathrm{Ox}} = V, \qquad \mathbf{x} \in \Gamma_G^2.$$

A typical geometry modelling an MOS transistor in two space dimensions is of the form

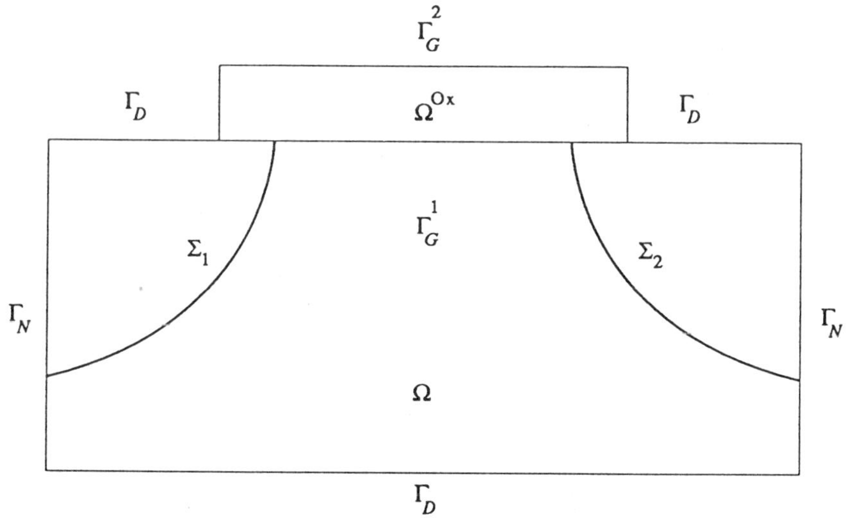

FIGURE 1

The curves Σ_1 and Σ_2 denote the *p*-*n* junctions. The regions where $C(x) > 0$ and $C(x) < 0$ hold are called the n and the p regions. Σ_1 and Σ_2 separate the n and p regions.

In Section 2 we scale (1.1) appropriately.

In Section 3 we carry out a singular perturbation analysis for the system (1.1). The geometry used corresponds to a two-dimensional MOS transistor. Particular emphasis is given to the occurrence of various time scales. The structure of this stiffness is analyzed in detail and it is determined which components of the solution evolve on the fast scale.

In Section 4 we present a difference scheme which takes into account this structure.

In Section 5 we give some numerical examples in order to demonstrate the results of the previous sections.

2. Scaling. Since the parameters in (1.1) are of different orders of magnitude, it is extremely important to scale (1.1) appropriately in order to obtain qualitative information. The scaling which seems to be the most useful for this purpose is of the form

$$(2.1)\qquad \begin{aligned} &t = l^2 D^{-1} t_s, \qquad x = l x_s, \qquad \psi = u_T \psi_s, \qquad n = \overline{C} n_s, \\ &p = \overline{C} p_s, \qquad \mathcal{T}_n = q\overline{C} D l^{-1} \mathcal{T}_{n_s}, \qquad \mathcal{T}_p = q\overline{C} D l^{-1} \mathcal{T}_{p_s}. \end{aligned}$$

Here $\overline{C} = \max\{|C(\mathbf{x})| : \mathbf{x} \in \Omega\}$ holds; l is the diameter of Ω and D is some average value of the diffusion coefficients D_n and D_p.

After scaling (1.1) assumes the form

$$(2.2)\qquad \begin{aligned} &\text{(a)}\quad -\lambda^2 \Delta \psi_s + n_s - p_s - C_s = 0, \\ &\text{(b)}\quad \frac{\partial}{\partial t_s} n_s - \operatorname{div} \mathcal{T}_{n_s} + R_s = 0, \\ &\text{(c)}\quad \frac{\partial}{\partial t_s} p_s + \operatorname{div} \mathcal{T}_{p_s} + R_s = 0, \\ &\text{(d)}\quad \mathcal{T}_{n_s} = D_{n_s}(\nabla n_s - n_s \nabla \psi_s), \\ &\text{(e)}\quad \mathcal{T}_{p_s} = D_{p_s}(-\nabla p_s - p_s \nabla \psi_x), \end{aligned}$$

with $C_s = C/\overline{C}$, $D_{n_s} = D_n/D$, $D_{p_s} = D_p/D$. If the Shockley–Read–Hall term is taken for the recombination rate R, R_s becomes

$$(2.3)\qquad R_s = K(np - \delta^4)(n + p + 2\delta^2)^{-1}$$

with $\delta = \sqrt{n_i/\overline{C}}$ and K an $O(1)$ constant. λ in (2.2)(a) is the scaled Debye length:

$$(2.4)\qquad \lambda = (\varepsilon U_T)^{1/2}(q l^2 \overline{C})^{-1/2}.$$

For the applications mentioned in the previous section, $\lambda = O(10^{-3})$ holds. For notational convenience we will omit the subscripts from now on.

After scaling, the diameter of Ω is $O(1)$. D, the average diffusion coefficient, will lie between 10 and 100. Thus, because of (2.1), the time scale considered for a 1μ device will be on the order of nanoseconds. The oxide layer Ω^{Ox} will usually be very thin compared to Ω. In this scaling its thickness will be of the same order of magnitude as λ. Although this has no physical foundation, we will set it proportional to λ

in order to capture the inversion layer effect. Thus, if in Figure 1, Γ_G^1 is located at $y = 0$ we assume that Γ_G^2 is at $y = d\lambda$ for some $d > 0$. The boundary conditions in scaled form read:

$$(2.5)\quad \begin{aligned}
&\text{(a)}\quad \psi = V + \psi_{\text{bi}}, \qquad n = n_b, \qquad p = p_b, \qquad \mathbf{x} \in \Gamma_D,\\
&\text{(b)}\quad \nabla\psi \cdot \mathbf{n} = \nabla n \cdot \mathbf{n} = \nabla p \cdot \mathbf{n} = 0, \qquad \mathbf{x} \in \Gamma_N,\\
&\text{(c)}\quad \psi = \psi^{\text{Ox}}, \qquad \partial_y \psi = \omega \partial_y \psi^{\text{Ox}},\\
&\text{(d)}\quad \mathcal{T}_n^y = \mathcal{T}_p^y = 0, \qquad \mathbf{x} \in \Gamma_G^1,\\
&\text{(e)}\quad \psi = V, \qquad \mathbf{x} \in \Gamma_G^2,\\
&\text{(f)}\quad \psi_{\text{bi}} = \ln[(C + \sqrt{C^2 + 4\delta^4})/(2\delta^2)].
\end{aligned}$$

So far, this scaling (or minor variations of it) has proven the most useful for analyzing the qualitative behavior of solutions. The main tool for such an analysis so far has been asymptotics for $\lambda \to 0$. This analysis has been performed in [**3**], [**4**] for the steady state case and in [**7**], [**8**] for the transient case in one space dimension. A singular perturbation approach of a slightly different flavor has been used in [**1**]. As mentioned previously, C will vary rapidly across Σ. In order to simplify the analysis in the next section, we will make three assumptions:

(2.6) *Assumption*: C is discontinuous across Σ,

(2.7) *Assumption*: $\nabla C \cdot \mathbf{n} = 0$ holds on Γ_N,

(2.8) *Assumption*: D_n and D_p are functions of $\mathbf{x}$ only.

3. Singular perturbation analysis. We now proceed to derive an asymptotic expression up to $O(\lambda)$ terms for the solution of (2.2). From [**3**] we know that the initial conditions, which can be assumed to be a steady state solution, are of the following form. Away from p-n junctions and the gate Γ_G they will, for $\lambda \to 0$, converge to the outer solution $\widetilde{w}^I = (\widetilde{\psi}^I, \widetilde{n}^I, \widetilde{p}^I, \widetilde{\mathcal{T}_n}^I, \widetilde{\mathcal{T}_p}^I)^{\mathrm{T}}$. Close to the p-n junctions Σ the steady state solution will be given, up to $O(\lambda)$ terms, by $\widehat{w}^I(\xi, s)$. (ξ, s) are the local layer variables given by the transformation

$$(3.1)\qquad x = a(s) + \lambda\xi\dot{b}(s), \qquad y = b(s) - \lambda\xi\dot{a}(s),$$

where ($x = a(s)$, $y = b(s)$) is a parametrization of the p-n junction curve Σ with $\dot{a}^2 + \dot{b}^2 = 1$. Close to the oxide-semiconductor interface $\Gamma_G^1(y = 0)_w^I$ will be given by the term $\overline{w}^I(x, \eta)$ with $\eta = y/\lambda$. We will therefore derive an asymptotic expansion of the same structure:

$$(3.2)\qquad w(x, y, t, \lambda) = \begin{cases} \widehat{w}(\xi, s, t) + O(\lambda) & \text{near } \Sigma,\\ \overline{w}(x, \eta, t) + O(\lambda) & \text{near } \Gamma_G^2,\\ \widetilde{w}(x, y, t) + O(\lambda) & \text{else.} \end{cases}$$

Let us first give the corresponding equations. Letting λ go to zero in (2.2) gives the system

$$
(3.3)\qquad
\begin{aligned}
&\text{(a)}\quad \tilde{n} - \tilde{p} - C = 0,\\
&\text{(b)}\quad \partial_t \tilde{n} - \operatorname{div} \widetilde{\mathcal{J}_n} + R(\tilde{n}, \tilde{p}) = 0,\\
&\text{(c)}\quad \partial_t \tilde{p} + \operatorname{div} \widetilde{\mathcal{J}_p} + R(\tilde{n}, \tilde{p}) = 0,\\
&\text{(d)}\quad \widetilde{\mathcal{J}_n} = D_n(\nabla \tilde{n} - \tilde{n} \nabla \widetilde{\psi}),\\
&\text{(e)}\quad \widetilde{\mathcal{J}_p} = D_p(\nabla \tilde{p} + \tilde{p} \nabla \widetilde{\psi}),
\end{aligned}
$$

for $\widetilde{w}(x, y, t) = (\widetilde{\psi}, \tilde{n}, \tilde{p}, \widetilde{\mathcal{J}_n}, \widetilde{\mathcal{J}_p})^{\mathrm{T}}$. Since C is discontinuous at Σ, (3.3) cannot be valid there because of (3.3)(a). Moreover, we need four conditions there to obtain a closed system. On the other hand, (3.3)(a) is compatible with the boundary conditions (2.5)(a), (b) at Γ_D and Γ_N. So we impose the boundary conditions:

(3.4)

$$
\begin{aligned}
&\text{(a)}\quad \widetilde{\psi} = V + \psi_{\mathrm{bi}}, \qquad \tilde{p} = p_b \quad \text{for } \mathbf{x} \in \Gamma_D \ (\rightarrow \tilde{n} = p_b + C = n_b),\\
&\text{(b)}\quad \nabla \widetilde{\psi} \cdot \mathbf{n} = \nabla \tilde{n} \cdot \mathbf{n} = 0 \quad \text{for } \mathbf{x} \in \Gamma_N \ (\rightarrow \nabla \tilde{n} \cdot \mathbf{n} = \nabla C \cdot \mathbf{n} = 0).
\end{aligned}
$$

The four conditions holding across the discontinuity at Σ are found by solving the layer equations and matching. Transforming (2.2) in the (ξ, s, t) variable set gives

$$
(3.5)\qquad
\begin{aligned}
&\text{(a)}\quad -\partial_\xi^2 \widehat{\psi} + \hat{n} - \hat{p} - C = O(\lambda),\\
&\text{(b)}\quad \partial_\xi(\widehat{\mathcal{J}_n} \cdot \mathbf{n}) = O(\lambda), \qquad \text{(c)}\quad \partial_\xi(\widehat{\mathcal{J}_p} \cdot \mathbf{n}) = O(\lambda),\\
&\text{(d)}\quad \partial_\xi \hat{n} - \hat{n} \partial_\xi \widehat{\psi} = O(\lambda), \qquad \text{(e)}\quad -\partial_\xi \hat{p} - \hat{p} \partial_\xi \widehat{\psi} = O(\lambda),
\end{aligned}
$$

where $\mathbf{n}(s) = (\dot{b}(s), -\dot{a}(s))^{\mathrm{T}}$ denotes the normal vector on Σ. For C in (3.5)(a) the one-sided limit at Σ has to be taken dependent on the sign of ξ. The tangential components of $\mathcal{J}_n$ and $\mathcal{J}_p$ are given by

$$
(3.6)\qquad
\begin{aligned}
&\text{(a)}\quad \widehat{\mathcal{J}_n} \cdot \mathbf{t} = D_n[\partial_s \hat{n} - \hat{n} \partial_s \widehat{\psi}],\\
&\text{(b)}\quad \widehat{\mathcal{J}_p} \cdot \mathbf{t} = D_p[-\partial_s \hat{p} - \hat{p} \partial_s \widehat{\psi}], \qquad \mathbf{t}(s) = (\dot{a}(s), \dot{b}(s))^{\mathrm{T}};
\end{aligned}
$$

letting λ go to zero in (3.5) and matching with the outer solution gives

$$
(3.7)\qquad
\begin{aligned}
&\text{(a)}\quad \widehat{\mathcal{J}_n}(\xi, s) \cdot \mathbf{n}(s) = \widetilde{\mathcal{J}_n}(0\pm, s),\\
&\text{(b)}\quad \widehat{\mathcal{J}_p}(\xi, s) \cdot \mathbf{n}(s) = \widetilde{\mathcal{J}_p}(0\pm, s),\\
&\text{(c)}\quad \hat{n}(\xi, s) = (n e^{-\widetilde{\psi}})(0\pm, s) e^{\widehat{\psi}(\xi, s)},\\
&\text{(d)}\quad \hat{p}(\xi, s) = (\tilde{p} e^{\widetilde{\psi}})(0\pm, s) e^{-\widehat{\psi}(\xi, s)},
\end{aligned}
$$

where $\widetilde{w}(0\pm, s)$ denotes the one-sided limits at Σ of $\widetilde{w} = (\psi, n, p, \mathcal{J}_n, \mathcal{J}_p)^{\mathrm{T}}$. Since $\widehat{w}$ has to be continuous at $\xi = 0$, we obtain the necessary conditions for $\widetilde{w}$ at Σ. They are that

$$(3.8) \qquad \widetilde{\mathcal{J}_n} \cdot \mathbf{n}, \qquad \widetilde{\mathcal{J}_p} \cdot \mathbf{n}, \qquad \tilde{n} \exp(-\widetilde{\psi}), \qquad \tilde{p} \exp(\widetilde{\psi})$$

stay continuous across Σ. The layer solution for the potential is obtained by solving (3.5)(a) using (3.7):

$$(3.9) \qquad \begin{gathered} -\partial_\xi^2 \widehat{\psi} + \tilde{n} \exp(\widehat{\psi} - \widetilde{\psi}) - \tilde{p} \exp(\widetilde{\psi} - \widehat{\psi}) - C = 0, \\ \widehat{\psi}(-\infty, s, t) = \widetilde{\psi}(0-, s, t), \qquad \widehat{\psi}(\infty, s, t) = \widetilde{\psi}(0+, s, t), \end{gathered}$$

where $\widetilde{\psi}, \tilde{n}, \tilde{p}$, and C are evaluated at the junction $\Sigma \cdot \widehat{\mathcal{J}_n} \cdot \mathbf{n}$, $\widehat{\mathcal{J}_p} \cdot \mathbf{n}$; $\hat{n}$ and $\hat{p}$ are then given by (3.7). $\widehat{\mathcal{J}_n} \cdot \mathbf{t}$, $\widehat{\mathcal{J}_p} \cdot \mathbf{t}$ are given by (3.6). Thus, after solving for the outer solution, one has to solve only (3.9) in order to obtain the solution near the p-n junction Σ. No layers occur in the normal components of the currents ($\mathcal{J}_n \cdot \mathbf{n}, \mathcal{J}_p \cdot \mathbf{n}$) and in the variables $n \exp(-\psi)$, $p \exp(\psi)$, which, in our scaling, are just the exponentials of the quasi-Fermi levels. This approximation procedure has to be applied near both of the p-n junctions Σ_1 and Σ_2 in Figure 1. Near the semiconductor-oxide interface Γ_G^2 the layer solution $\widetilde{w}(x, y, t)(y = y/\lambda, \overline{w} = (\psi, \overline{n}, \overline{p}, \overline{\mathcal{J}}_n, \overline{\mathcal{J}}_p)^{\mathrm{T}}$ obeys

(3.10)

$$\text{(a)} \quad -\partial_\eta^2 \overline{\psi} + \overline{n} - \overline{p} - C(x, 0) = 0,$$

$$\text{(b)} \quad \partial_\eta \overline{\mathcal{J}}_n^y = O(\lambda), \qquad \text{(c)} \quad \partial_\eta \overline{\mathcal{J}}_p^y = O(\lambda),$$

$$\text{(d)} \quad \partial_\eta \overline{n} - \overline{n} \partial_\eta \overline{\psi} = O(\lambda), \qquad \text{(e)} \quad \partial_\eta \overline{p} + \overline{p} \partial_\eta \overline{\psi} = O(\lambda),$$

$$\text{(f)} \quad \overline{\mathcal{J}}_n^x = D_n(\partial_x n - n \partial_x \psi), \qquad \text{(g)} \quad \overline{\mathcal{J}}_p^x = D_p(-\partial_x \overline{p} - \overline{p} \partial_x \overline{\psi}).$$

In the oxide

$$(3.10) \qquad \text{(h)} \quad \partial_\eta^2 \psi^{\mathrm{Ox}} = O(\lambda)$$

holds. AT Γ_G^1 and Γ_G^2 the interface and boundary conditions read

(3.11)

$$\text{(a)} \quad \overline{\psi}(x, 0, t) = \psi^{\mathrm{Ox}}(x, 0, t), \qquad \partial_\eta \overline{\psi}(x, 0, t) = \omega \partial_\eta \overline{\psi}^{\mathrm{Ox}}(x, 0, t),$$

$$\text{(b)} \quad \overline{\mathcal{J}}_n^y(x, 0, t) = \overline{\mathcal{J}}_p^y(x, 0, t) = 0,$$

$$\text{(c)} \quad \psi^{\mathrm{Ox}}(x, d, t) = V.$$

Again, letting λ go to zero and matching with $\widetilde{w}$ gives

$$
(3.12)\quad
\begin{aligned}
&\text{(a)}\quad \overline{\mathcal{T}}_n^y(x,\eta,t) \equiv \widetilde{\mathcal{T}}_n^y(x,0,t),\\
&\text{(b)}\quad \overline{\mathcal{T}}_p^y(x,\eta,t) = \widetilde{\mathcal{T}}^y(x,0,t),\\
&\text{(c)}\quad \overline{n}(x,\eta,t) = (\tilde{n}e^{-\widetilde{\psi}})(x,0,t)\exp\overline{\psi}(x,\eta,t),\\
&\text{(d)}\quad \overline{p}(x,\eta,t) = (\tilde{p}e^{\widetilde{\psi}})(x,0,t)\exp(-\overline{\psi}(x,\eta,t)).
\end{aligned}
$$

For ψ and ψ^{Ox} the equations read, after inserting (3.12) for $\overline{n}$ and $\overline{p}$,

$$
(3.13)\quad
\begin{aligned}
&\text{(a)}\quad -\partial_\eta^2\overline{\psi} + \tilde{n}\exp(\overline{\psi}-\widetilde{\psi}) - \tilde{p}\exp(\widetilde{\psi}-\overline{\psi}) - C = 0,\\
&\text{(b)}\quad \overline{\psi}(x,0,t) + \frac{\omega}{d}\partial_\eta\overline{\psi}(x,0,t) = V, \quad \overline{\psi}(x,\infty,t) = \widetilde{\psi}(x,0,t),\\
&\text{(c)}\quad \psi^{\mathrm{Ox}} = V + \omega^{-1}\partial_\eta\overline{\psi}(x,0,t)(\eta - d).
\end{aligned}
$$

Thus, after solving for $\widetilde{w}$, one, again, only has to solve the nonlinear two-point boundary value problem (3.13). $\overline{n}$, $\overline{p}$, $\overline{\mathcal{T}}_n^y$, $\overline{\mathcal{T}}_p^y$ are then given by (3.12). $\overline{\mathcal{T}}_n^x$, $\overline{\mathcal{T}}_p^x$ are given by

$$
(3.14)\qquad \overline{\mathcal{T}}_n^x = D_n(\partial_x\overline{n} - n\partial_x\overline{\psi}), \qquad \overline{\mathcal{T}}_p^x = D_p(-\partial_x\overline{p} - \overline{p}\partial_x\overline{\psi}).
$$

After inserting (3.12) for $\overline{n}$ and $\overline{p}$ one obtains

$$
(3.15)\qquad \overline{\mathcal{T}}_n^x = \widetilde{\mathcal{T}}_n^x\exp(\overline{\psi}-\widetilde{\psi}), \qquad \overline{\mathcal{T}}_p^x = \widetilde{\mathcal{T}}_p^x\exp(\widetilde{\psi}-\overline{\psi}).
$$

This leads to the following procedure for obtaining a solution on the t-timescale.

Step 1: Solve the reduced problem consisting of the equations (3.3) and the boundary conditions (3.4), (2.5)(d).

Step 2: Solve the two O.D.E. boundary value problems (3.9) and (3.13) for $\widehat{\psi}$ and $\overline{\psi}$.

Step 3: Compute the rest of the solution in the layers by virtue of (3.7), (3.12).

(3.16) REMARK. The only problem requiring initial conditions is the reduced problem (3.3), (3.4).

For the initial functions n and p^I to be I compatible with the reduced problem, they have to satisfy

$$
(3.17)\qquad \text{(a)}\quad n^I - p^I - C = O(\lambda), \qquad \text{(b)}\quad \operatorname{div}(\mathcal{T}_n^I + \mathcal{T}_p^I) = O(\lambda),
$$

away from layer regions. It is reasonable to assume (3.17)(a) to hold, because otherwise one would prescribe a field $\nabla\psi$ at time $t=0$, which is uniformly $O(1/\lambda^2)$ in the whole of Ω because of (2.2)(a). (3.17)(b) will hold if we prescribe initial conditions which are steady state solutions.

Let us now investigate the behavior of the solution if we do not start with a steady state and (3.17)(b) is violated. This introduces a fast time scale. As shown in [7] the appropriate scale is $O(\lambda^2)$ which corresponds to the dielectric relaxation time. Introducing the new variable $\tau = t/\lambda^2$ in (2.2) gives

$$\text{(3.18)}\qquad \begin{aligned} &\text{(a)} && -\lambda^2 \Delta\psi + n - p - C = 0,\\ &\text{(b)} && \partial_\tau n - \lambda^2 \operatorname{div} \mathcal{J}_n + \lambda^2 R = 0,\\ &\text{(c)} && \partial_\tau p + \lambda^2 \operatorname{div} \mathcal{J}_p + \lambda^2 R = 0,\\ &\text{(d)} && \mathcal{J}_n = D_n(\nabla n - n\nabla\psi),\\ &\text{(e)} && \mathcal{J}_p = D_p(-\nabla p - p\nabla\psi). \end{aligned}$$

If we assume that the currents stay bounded, $\lambda \to 0$ gives

$$\text{(3.19)}\qquad \partial_\tau n = \partial_\tau p = 0 \to n = n^I,\ p = p^I.$$

The equation for ψ on the fast time scale is obtained by differentiating (3.18)(a) w.r.t. time and inserting from (3.18)(b), (c). Denoting ψ, $\mathcal{J}_n$, $\mathcal{J}_p$ on the fast time scale $\tau = t/\lambda^2$ by ϕ, I_n, I_p we obtain

$$\text{(3.20)}\qquad \begin{aligned} &\text{(a)} && -\partial_\tau \Delta\phi + \operatorname{div}(I_n + I_p) = 0,\\ &\text{(b)} && I_n + I_p = D_n \nabla n^I - D_p \nabla p^I - (D_n n^I + D_p p^I)\nabla\phi. \end{aligned}$$

The boundary conditions for ϕ remain the same as for ψ:

$$\text{(3.21)}\qquad \begin{aligned} &\text{(a)} && \phi = V, && \mathbf{x} \in \Gamma_D,\\ &\text{(b)} && \nabla\phi \cdot \mathbf{n} = 0, && \mathbf{x} \in \Gamma_N,\\ &\text{(c)} && \phi = \psi^{\mathrm{Ox}}(t = 0), && \mathbf{x} \in \Gamma_G^2. \end{aligned}$$

The initial conditions for ϕ are obtained by solving Poisson's equation at $t = 0$:

$$\text{(3.22)}\qquad \begin{aligned} &\text{(a)} && \phi(x, y, t = 0) = \psi^I(x, y),\\ &\text{(b)} && -\lambda^2 \Delta\psi^I + n^I - p^I - C = 0. \end{aligned}$$

If ϕ in (3.20) reaches a steady state, the divergence of the drift and diffusion vanishes ($\operatorname{div}(I_n + I_p) \to 0$), and this solution can be matched to the expression for the slow time scale derived in steps (1)–(3) at $t = 0$. Thus, for arbitrary initial data, which are such that they produce a bounded potential and current densities at $t = 0$, one first solves (3.20) for ϕ on the fast time scale. This means that the field is adjusted on the scale of the dielectric relaxation time to produce $\operatorname{div}(\mathcal{J}_n + \mathcal{J}_p) = 0$. Then the slow time scale solution is obtained by performing Steps (1)–(3).

4. Time differencing. When analyzing stability properties of time discretization methods for (2.2), it is necessary to know the maximal size of the real parts of the eigenvalues of the corresponding eigenvalue problem in order to obtain meaningful bounds on the time step. In **[5]**, **[6]**, the corresponding linearized eigenvalue problem has been analyzed. Linearizing (2.2) around some solution (ψ, n, p) in the direction (ϕ, u, v) gives

$$(4.1)\qquad \begin{aligned} &\text{(a)} \quad -\lambda^2\Delta\phi + u - v = 0,\\ &\text{(b)} \quad -\partial_t u + \operatorname{div} I_n + au + bv = 0,\\ &\text{(c)} \quad -\partial_t v - \operatorname{div} I_p + au + bv = 0,\\ &\text{(d)} \quad I_n = D_n(\nabla u - u\nabla\psi - n\nabla\phi),\\ &\text{(e)} \quad I_p = D_p(-\nabla v - v\nabla\psi - p\nabla\phi).\end{aligned}$$

In the corresponding eigenvalue problem (4.1)(b), (c) are replaced by

$$(4.1)\qquad \begin{aligned} &\text{(f)} \quad -\omega u + \operatorname{div} I_n + au + bv = 0,\\ &\text{(g)} \quad -\omega v - \operatorname{div} I_p + au + bv = 0,\end{aligned}$$

with ω the eigenvalue. In **[6]** the eigenvalue problem (4.1) has been analyzed for the case when ψ, n, and p are equilibrium solutions, and it has been shown that ω has to be real and negative in this case. In **[5]** the linearization around a general solution has been treated. In general, the corresponding linear operator generates an analytic semigroup and

$$(4.2)\qquad \operatorname{Re}\omega \le K[\|\nabla \ln(n\exp(-\psi))\|_\infty + \|\nabla \ln(p\exp(\psi))\|_\infty]$$

holds where K is independent of λ. Since it has been shown in the previous section that $n\exp(-\psi)$ and $p\exp(\psi)$ do not exhibit any layers, the real part of ω is bounded by a constant independent of λ. A simple argument then gives linearized Von Neumann stability for most of the standard difference schemes if

$$(4.3)\qquad \Delta t \le \text{const}[\|\nabla(\ln(ne^{-\psi}))\|_\infty + \|\nabla(\ln(pe^{\psi}))\|_\infty$$

holds (see **[5]**).

The presence of various time scales complicates the choice of an appropriate time discretization. Discretizations which are just A-stable, like for instance Crank–Nicholson, will produce oscillations when the boundary conditions are varied too rapidly **[9]**. On the other hand, the asymptotic analysis outlined in the previous section provides us with an understanding of the structure of the stiffness, which can be used in the design of time differencing schemes. We know that the solution will vary essentially smoothly on the time scale of interest (t). Whenever the condition $\operatorname{div}(\mathscr{T}_n + \mathscr{T}_p) = O(\lambda^2)$ is violated, the potential ψ will be adjusted on the faster (t/λ^2) time scale. In [9] a time discretization based on the trapezoidal rule has been developed which takes this phenomenon into account:

$$(4.4)\quad \begin{aligned} &\text{(a)}\quad -\lambda^2\Delta\psi_{k+1} + n_{k+1} - p_{k+1} - C = 0,\\ &\text{(b)}\quad n_{k+1} - \frac{\Delta t}{4}[\operatorname{div}(3\mathscr{T}_{n_{k+1}} + \mathscr{T}_{p_{k+1}}) - 2R_{k+1}]\\ &\qquad = n_k - \frac{\Delta t}{4}[\operatorname{div}(\mathscr{T}_{p_k} - \mathscr{T}_{n_k}) + 2R_k],\\ &\text{(c)}\quad p_{k+1} - \frac{\Delta t}{4}[\operatorname{div}(-3\mathscr{T}_{p_{k+1}} - \mathscr{T}_{n_{k+1}}) - 2R_{k+1}]\\ &\qquad = p_k - \frac{\Delta t}{4}[\operatorname{div}(\mathscr{T}_{p_k} - \mathscr{T}_{n_k}) + 2R_k]. \end{aligned}$$

Here R_k denotes $R(n_k, p_k)$. To obtain the discrete equivalent of (3.20), one subtracts (4.4)(c) from (4.4)(b) and obtains after inserting from (4.4)(a)

$$(4.5)\qquad -\lambda^2\Delta\psi_{k+1} - \Delta t\operatorname{div}(\mathscr{T}_{n_{k+1}} + \mathscr{T}_{p_{k+1}}) = -\lambda^2\Delta\psi_k.$$

Thus, the fast time scale equation is implicitly discretized by the backward Euler method and it is assured that $\operatorname{div}(\mathscr{T}_n + \mathscr{T}_p) = O(\lambda)$ holds. To compute the local discretization error it is beneficial to rewrite (4.4)(b) into

$$(4.6)\quad \begin{aligned} n_{k+1} - \frac{\Delta t}{2}[\operatorname{div}\mathscr{T}_{n_{k+1}} + R_{k+1}] = n_k &+ \frac{\Delta t}{2}[\operatorname{div}\mathscr{T}_{n_k} + R_k]\\ &- \frac{\lambda^2\Delta t}{4}\Delta[\psi_{k+1} - 2\psi_k + \psi_{k-1}] \end{aligned}$$

using (4.4)(a). A similar expression holds for (4.4)(c). Thus, the local discretization error is of the form

$$(4.7)\qquad O(\Delta t^2) + O(\lambda^2\Delta t)$$

and the method is of first order formally and of second order whenever $\Delta t \gg \lambda^2$ holds.

5. Numerical results. In this section, we present some numerical examples to demonstrate the results of Sections 3 and 4. To show the validity of the asymptotic expansions, (2.2) has been solved for a simplified one-dimensional problem. Here the Dirichlet boundary conditions (2.5)(a) are imposed at $x = \pm 1$ and the junction Σ is located at $x = 0$. Figures 2 and 3 show the hole distribution for $x \in [0, 1]$, $t > 0$. The device has been switched from equilibrium to 0.5 V forward bias in 0.1

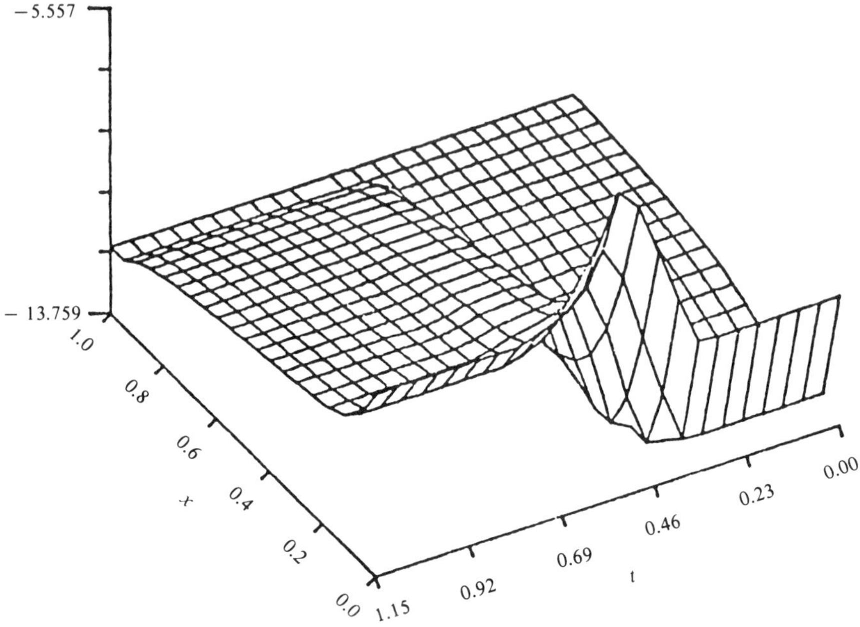

FIGURE 2

ns. Figure 2 shows the hole distribution according to (2.2). Figure 3 is a plot of the solution of the reduced problem (3.3). They coincide except for the spatial layer at $x = 0$. In both cases the steady state has been taken as initial datum ($\mathrm{div}(\mathcal{T}_n + \mathcal{T}_p) = 0$); so no initial layer occurs. In Figures 5 and 6 the initial datum has been perturbed. Figure 4 shows $\psi(x, t = 0)$ corresponding to this initial datum and the steady state. Figure 5 shows ψ on the τ time scale. Figure 6 shows the difference between the perturbed solution and the reduced problem. Table 3 demonstrates

TABLE 3

l	PSI	N	P	$\mathcal{T} = \mathcal{T}_n + \mathcal{T}_p$	$\mathcal{T}P$
0	0.33E − 01	0.89E + 00	0.15E + 00	0.58E + 00	0.39E + 00
1	0.10E − 01	0.17E + 00	0.46E − 01	0.18E + 00	0.12E + 00
2	0.16E − 02	0.27E − 01	0.74E − 02	0.29E − 01	0.20E − 01
3	0.17E − 02	0.62E − 02	0.77E − 03	0.31E − 02	0.20E − 02
4	0.39E − 04	0.16E − 02	0.18E − 03	0.80E − 03	0.45E − 03
5	0.84E − 05	0.38E − 03	0.38E − 04	0.21E − 03	0.94E − 04

the convergence rate of the scheme proposed in Section 4. Here the spatial mesh has been fixed and the time steps have been successively halved. $l = 0$ corresponds to the coarsest mesh. The table shows the errors, measured in the discrete L^2 norm between the solutions on the corresponding mesh and the "numerically exact" solution (for $l = 6$). All computations have been carried out using Scharfetter–Gummel finite differences in space.

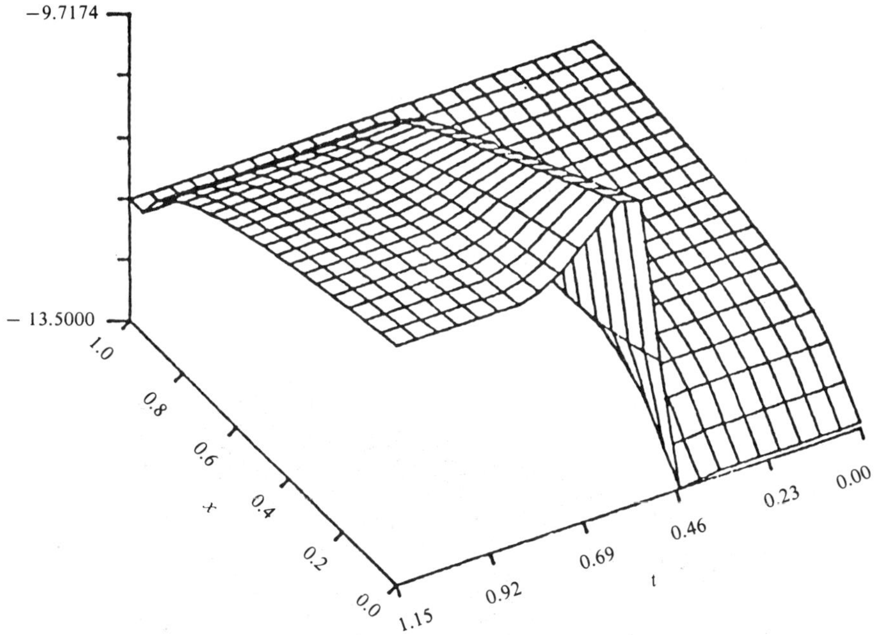

FIGURE 3

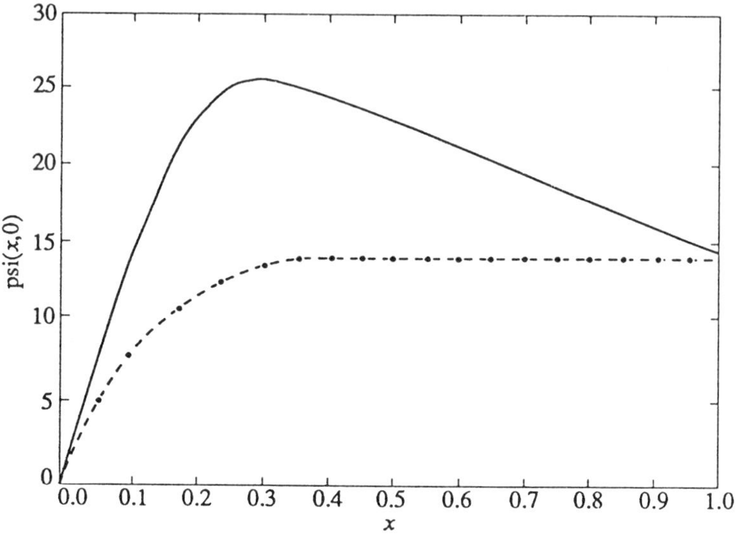

FIGURE 4

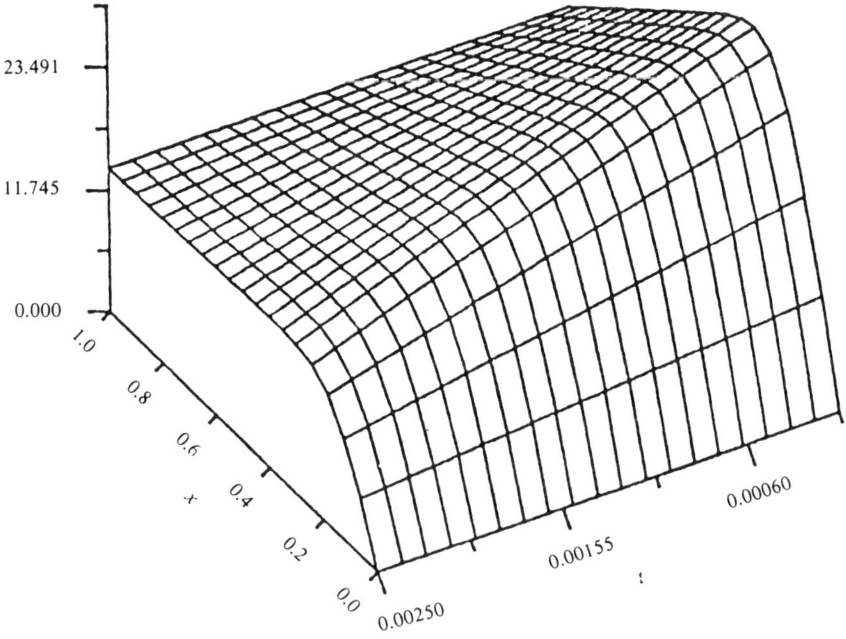

FIGURE 5

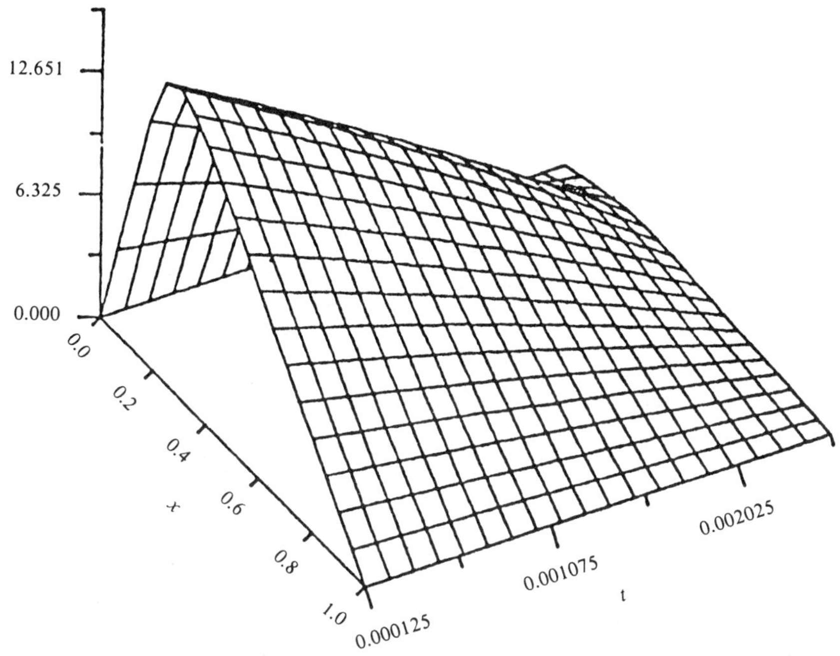

FIGURE 6

REFERENCES

1. F. Brezzi, *Singular perturbation analysis of (strongly) reverse biased semiconductor devices*, this volume.

2. P. Markowich, *The Stationary Semiconductor Equations*, Springer, 1985.

3. ———, *A singular perturbation analysis of the steady state semiconductor equations*, SIAM J. Appl. Math. **5**, No. 44, 896–928.

4. P. Markowich and C. Ringhofer, *A singularly perturbed boundary value problem modelling a semiconductor device*, SIAM J. Appl. Math. **44**, No. 2 (1984), 231–256.

5. ———, *The stability of the transient semiconductor equations*, ZAMM **67**, No. 7 (1987), 319–332.

6. M. Mock, *Analysis of Mathematical Models of Semiconductor Devices*, Boole Press, 1983.

7. C. Ringhofer, *An asymptotic analysis of a transient P-N junction model*, SIAM J. Appl. Math. **47**, No. 3 (1987), 624–642.

8. ———, *A singular perturbation analysis for the transient semiconductor equations*, IMA J. Appl. Math. **39** (1987), 17–32.

9. ———, *Difference methods for transient device simulation*, Proc. Septième Colloque International sur les Méthodes de Calcul Scientifique (R. Glowinski, ed.), Paris, 1985.

10. S. Selberherr, *Analysis and Simulation of Semiconductor Devices*, Springer, 1984.

11. W. Van Roosbroeck, *Theory of flow of electrons and holes in semiconductors*, Bell Syst. Tech. J. **29** (1950), 560–607.

ARIZONA STATE UNIVERSITY, TEMPE, ARIZONA 85287

Lectures in Applied Mathematics
Volume **25** (1990)

Singular Perturbation Analysis of (Strongly) Reverse Biased Semiconductor Devices

F. BREZZI

1. Introduction. To fix the notation let us first consider the equations that rule the behaviour of a (silicon) semiconductor device in their general form **[12]**

$$(1)\qquad \begin{cases} \operatorname{div}(\varepsilon \nabla \psi) = -q(D - n + p), \\ \dfrac{\partial p}{\partial t} = q \operatorname{div}(D_p \nabla p + \mu_p p \nabla \psi) + R, \\ \dfrac{\partial n}{\partial t} = q \operatorname{div}(D_n \nabla n - \mu_n n \nabla \psi) + R \\ \quad + \text{ initial and boundary conditions.} \end{cases}$$

As usual, the unknown functions are ψ = electric potential and p, n = carrier's concentrations (for holes and electrons respectively). $D(x)$ is the doping profile (known), ε the permittivity, D_p, μ_p, D_n, μ_n the diffusivity and mobility coefficients (for holes and electrons respectively) and R is the recombination-generation term. If we take, for simplicity, a one-dimensional strongly reverse biased device, with D constant in each region, the stationary solution can be roughly described as follows: the device $[a, b]$ can be split in three parts $P_1 \cup P_2 \cup P_3$ where

$$\begin{aligned} &\text{in } P_1: p \simeq -D; \quad \psi \simeq \psi(a); \quad n \simeq 0; \\ &\text{in } P_2: p \simeq 0; \quad \varepsilon\psi'' \simeq -qD; \quad n \simeq 0; \\ &\text{in } P_3: p \simeq 0; \quad \psi \simeq \psi(b); \quad n \simeq D. \end{aligned}$$

1980 *Mathematics Subject Classification* (1985 *Revision*). Primary 78A30, 34B15.
Partially supported by M.P.I. 60%.

In such cases the solution can be computed in a simpler way, by means of the so-called "Total Depletion Assumption" which roughly amounts to assuming that

$$p = -D\chi_{\{x|\psi(x)=\psi(a)\}},$$
$$n = D\chi_{\{x|\psi(x)=\psi(b)\}},$$

where $\chi_E(x)$ = characteristic function of the set $E := 1$ (if $x \in E$) or 0 (if $x \notin E$). We are therefore left with a single nonlinear equation

$$\text{(2)} \qquad \begin{cases} \operatorname{div}(\varepsilon \nabla \psi) = -qD(1 - \chi_{\{\psi=\psi_a\}} - \chi_{\{\psi=\psi_b\}}) \\ \quad + \text{ boundary conditions} \end{cases}$$

which, in its turn, can then be written as a *double obstacle problem* and efficiently solved as a *variational inequality* (see [7], [9]).

Here, in some simplified cases, we shall present the T.D.A. as the limit problem (as a suitable parameter λ goes to zero) of the original problem (1), seen as a singular perturbation problem. We shall also present a uniqueness theorem for (1), valid for λ small enough. The uniqueness of the solution of (2) (properly rewritten as a variational inequality) is obvious. We end the presentation with a few remarks on the numerical treatment of (1). Again, the numerical treatment of (2) is a standard problem (see, in the context of semiconductor devices, [7], [9] and then [8] for an easier handling of mixed methods). The material presented here comes mainly from the joint papers [1], [2], [3], [5].

2. The simplified case and the scaling. For the sake of simplicity:

(1) we consider a one-dimensional device $[a, b]$;
(2) we look for the stationary solution;
(3) we neglect the recombination-generation terms;
(4) we assume ε, D_p, μ_p, D_n, μ_n to be constant and we assume the Einstein relation

$$D_p/\mu_p = D_n/\mu_n = kT/q$$

(k = Boltzmann constant, T = absolute temperature);

(5) we assume $D(x)$ to be of the form

$$D(x) = D_0 \operatorname{sign}(x) \qquad (D_0 > 0).$$

Note that, implicitly, we assumed $ab < 0$. The use of two different constant values for $D(x)$ (one in each region) will only complicate the notation a bit but will not change the validity of the results.

The simplified problem becomes now:

$$(3)\qquad \begin{cases} \varepsilon\psi'' = -q(D_0\,\mathrm{sign}(x) - n + p) \text{ in }]a,b[, \\ ((D_n/\mu_n)n' - n\psi')' = 0 \text{ in }]a,b[, \\ ((D_p/\mu_p)p' + p\psi')' = 0 \text{ in }]a,b[\end{cases}$$

with the boundary conditions

$$(4)\qquad \begin{cases} \psi(a) = -\overline{\psi}; & \psi(b) = \overline{\psi}, \\ n(a) = D_0\beta; & n(b) = D_0(1+\beta), \\ p(a) = D_0(1+\beta); & p(b) = D_0\beta, \end{cases}$$

where

$$(5)\qquad \begin{cases} 2\overline{\psi} = V + \frac{kT}{q}\lg\left(\frac{1+\beta}{\beta}\right), \\ \beta(1+\beta) = n_i^2/D_0^2 \qquad (\beta > 0) \end{cases}$$

with V = applied potential and n_i = intrinsic concentration.

The scaling that we adopted is

$$(6)\qquad \begin{cases} \psi_s = \frac{\varepsilon}{qD_0d^2}; & D_s = \frac{D}{D_0} = \mathrm{sign}(x), \\ p_s = p/D_0; & n_s = n/D_0, \end{cases}$$

where d = length of the device, which is therefore scaled to $1: b_s - a_s = 1$. With the scaling (6) the problem (3), (4), (5) becomes

$$(7)\qquad \begin{cases} \psi_s'' = -D_s + n_s - p_s; & \psi_s(a_s) = -\overline{\psi}_s; & \psi_s(b_s) = \overline{\psi}_s, \\ (\lambda p_s' + p_s\psi_s')' = 0; & p_s(a_s) = 1+\beta; & p_s(b_s) = \beta, \\ (\lambda n_s' - n_s\psi_s')' = 0; & n_s(a_s) = \beta; & n_s(b_s) = 1+\beta, \end{cases}$$

where

$$(8)\qquad \lambda = \frac{\varepsilon kT}{q^2D_0d^2},$$

$$(9)\qquad 2\overline{\psi}_s = \frac{\varepsilon}{qD_0d^2}\left(V + \frac{kT}{q}\lg\left(\frac{1+\beta}{\beta}\right)\right).$$

REMARK 1. Other scalings are often used in the literature. See for instance the lectures of P. A. Markowich and C. A. Ringhofer at this workshop. It is clear that all of them are equivalent to each other, as long as you do not change the values of the physical and geometrical parameters (that is, *before* the singular perturbation analysis). The difference between one scaling and another is then (if no singular perturbation analysis is done) purely *psychological*. In this respect our preference for the scaling (6) is mainly dictated by the following two reasons: (1) with the scaling (6) ψ_s is a smooth function ($\|\psi_s\|_{L^\infty}$ is small and $\|\psi_s''\|_{L^\infty} \simeq 1$) and the corresponding equation (7.1) is "nice";

(2) the presence of λ in the "continuity equations" (7.2) and (7.3) emphasizes the fact that they are *advection dominated.* Note that this is not an *effect* of the scaling: the advection *is* physically dominant, in most applications, at least in some part of the domain (here, for instance, in the depleted region P_2).

REMARK 2. It is easy to check that, in most of the cases, the value of β given by (5.2) is very small. Thus, from the practical point of view, one can change the boundary conditions in (7) to

$$p_s(a_s) = n_s(b_s) = 1; \qquad p_s(b_s) = n_s(a_s) = 0. \tag{10}$$

This is, in a sense, totally justified in our case, since the relative change of the solution is negligible. However, it is of some interest to see that the *qualitative behaviour* of the solution undergoes a drastic change. For instance, in the interval $[a_s, 0]$ one has (see **[2]**, **[3]**): with the original boundary conditions as in (7), ψ_s is increasing, p_s is decreasing, and n_s can have at most one minimum; with the modified boundary conditions (10), instead, n_s is increasing, p_s can have at most one maximum and ψ_s can have (and generally does) at most one minimum. To understand how one can go from one situation to the other with a very minor change, remember that in the region P_1 we have ψ_s, p_s, and n_s practically constant.

3. The singular perturbation analysis. We want now to study the limit behaviour of (7) as λ goes to zero. At this point the difference among the various scalings stops being a psychological question and becomes a mathematical one. The reason for that is the following: one has to decide which of the six quantities appearing in the expression of λ (see (8)) is responsible for the fact that λ goes to zero, and change the boundary conditions accordingly. (Actually, it is clear that one has more than six possible choices, but let us stick to the six simplest ones, for the sake of simplicity.)

If one wants to be able to write the limit problem for the original unscaled variables ψ, p, and n, it seems reasonable to leave fixed all the quantities that enter the scaling (6). Hence, a reasonable respect for the Boltzmann constant leaves us with the temperature going to zero. However, it has to be pointed out that this is just some mathematical trick in order to have a better insight to the problem (7) itself and to the total depletion assumption. Coming back to our question, we have to make explicit the dependence on T of the boundary conditions. In view of (5) we look for the dependence of n_i^2 on T:

$$n_i^2 = 4(2\pi kT/h^2)^3(m_n m_p)^{3/2}e^{W_f/(kT)} \tag{11}$$

where h is the Plank constant, m_n and m_p are the effective masses (of electrons and holes respectively) and W_f is the (energy) width of the semiconductor forbidden band. Hence β as a function of λ behaves (for $T \to 0$) as

$$\beta \simeq \lambda^3 e^{-A/\lambda} \tag{12}$$

and for $\lambda \to 0$ we have

$$\beta \to 0; \quad 2\overline{\psi} \to V + W_f/q. \tag{13}$$

Note that W_f/q is a kind of built-in potential at 0° Kelvin.

We are now able to state the convergence theorem for $\lambda \to 0$.

THEOREM 1. *Let, for $\lambda > 0$, $\psi_s(\lambda)$, $p_s(\lambda)$, $n_s(\lambda)$, be a solution of* (7). *Then, for $\lambda \to 0$ we have*

$$\psi_s(\lambda) \to \mathring{\psi}_s, \qquad p_s(\lambda) \to \mathring{p}_s, \qquad n_s(\lambda) \to \mathring{n}_s, \tag{14}$$

where $(\mathring{\psi}_s, \mathring{p}_s, \mathring{n}_s)$ satisfies:

$$\mathring{\psi}_s'' = -D_s + \mathring{n}_s - \mathring{p}_s; \qquad \mathring{\psi}_s(a_s) = -\overline{\mathring{\psi}}_s; \qquad \mathring{\psi}_s(b_s) = \overline{\mathring{\psi}}_s; \tag{15}$$

$$\mathring{\psi}_s \in C^1([a_s, b_s]); \tag{16}$$

$$\mathring{p}_s = \chi_{\{\mathring{\psi}_s = -\overline{\mathring{\psi}}_s\}}; \qquad \mathring{n}_s = \chi_{\{\mathring{\psi}_s = \overline{\mathring{\psi}}_s\}}; \tag{17}$$

$$2\overline{\mathring{\psi}}_s = \frac{\varepsilon}{qD_0 d^2}(V + W_f/q). \tag{18}$$

Note that it is still possible to write the limit problem (15)–(18) in terms of the original unscaled variables. Note in particular that the width of the depleted region will be, in the limit

$$\min\left\{2\left(\varepsilon\frac{(qV + W_f)}{q^2 D_0}\right)^{1/2}, d\right\} \tag{19}$$

which is, in some sense, independent of d for reasonable devices.

REMARK 3. The proof of Theorem 1, which can be found in [**1**], is based on the a priori bounds

$$\|p_s(\lambda)\|_{W^{1,1}} + \|n_s(\lambda)\|_{W^{1,1}} + \|\psi_s(\lambda)\|_{W^{2,\infty}} \leq C \quad \text{(indep. of } \lambda) \tag{20}$$

that should also be used for making (14) more precise. Previous results of this nature can be found in [**2**], [**3**], [**6**].

THEOREM 2. *There exists a* $\lambda_0 \geq 0$ *such that for all* $\lambda \in]0, \lambda_0]$ *the solution of* (7) *is unique.*

REMARK 4. The proof of Theorem 2 can be found in [**1**]. A similar result can also be obtained for the simplified boundary conditions (10); see [**3**]. Previous uniqueness results were known only under the assumptions: λ big enough or V small enough. See for instance [**10**], [**11**] and the references therein. □

4. A few numerical considerations. When solving (7) numerically, after a suitable discretization, one usually employs some kind of Newton method. It seems, out of our experience, that λ is indeed a very good continuation parameter to be used in conjunction with the Newton method itself. Actually, from one side it is known that the Newton method, in particular for problems of the type (7) for λ small, needs a starting point that is very close to the solution. On the other side it is easy to check that, when λ decreases, the sharp fronts in p and n become sharper and sharper, but basically their location changes very little. Therefore it is very convenient to start with an artificially big λ (and then (7) is easy) and then to decrease it slowly towards the actual physical value.

As far as the discretization itself is concerned, the equation (7.1) is "easy" and can be discretized, somehow, by any reasonable approach (finite differences, finite elements, etc). No particular care is required for it. On the contrary (7.2) and (7.3) are advection dominated and require much more attention. To discretize them in the original unknowns p and n is somehow risky, due to the presence of the zero-order terms ($p\psi''$ and $-n\psi''$) which have the *wrong sign* in half of the domain. Discretizing the Slotboom variables

$$\rho = pe^{(\psi+\overline{\psi})/\lambda}, \qquad \sigma = ne^{(\overline{\psi}-\psi)/\lambda} \tag{21}$$

(and then getting back to p and n before or after the solution of the linear system) seems more convenient (and most commonly done) due to the symmetry of the corresponding equations:

$$(e^{-(\psi+\overline{\psi})/\lambda}\rho')' = 0; \qquad (e^{(\psi-\overline{\psi})/\lambda}\sigma')' = 0. \tag{22}$$

The use of a mixed method for (22) is particularly interesting (see [**4**], [**5**]) since it gives back the widely used Scharfetter–Gummel scheme in one dimension. However, other methods borrowed from the theory of advection dominated problems can be profitably used.

As far as the error estimates are concerned, little is known on the coupled system (7). It can be shown that, for the unipolar case, the scheme

$$\delta_c^2 \psi_h = 1 - \rho_h e_M^{-\psi_h/\lambda}, \tag{23}$$

$$\left(\lambda + \gamma h + \frac{h|\delta\psi|}{2}\right)\delta_c^2 \rho h - \delta\psi_h \delta_c \rho_h = 0 \tag{24}$$

produces an error that can be estimated as

$$\|\psi - \psi_h\|_{H^1} \le c(h\|\psi\|_{H^2} + e^{-c_1\overline{\psi}/\lambda} + \varepsilon_M), \tag{25}$$

$$\|\rho - \rho_h\|_{L^1} \le ch\|\rho\|_{W^{1,1}}. \tag{26}$$

In (23)–(26) δ_c^2 is the usual centered second-order finite difference, δ the forward first-order f.d., δ_c the centered first-order f.d., e_M^x the exponential function as computed by the machine, $\gamma = \max|\delta\psi|$, c and c_1 positive constants independent of h and λ, and ε_M is the "zero-machine" (say, here, $\varepsilon_M = \sup_{x\ge 0} e^x - e_M^x$).

REFERENCES

1. F. Brezzi, A. Capelo, and L. Gastaldi, *A singular perturbation analysis for semiconductor devices*, SIAM J. Math Anal. **20** (1989), 372–387.

2. F. Brezzi, A. Capelo, and L. D. Marini, *Singular perturbation problems in semiconductor devices*, Lecture Notes in Math (J. P. Hennart, ed.), **1230**, Springer, 1986, pp. 191–198.

3. F. Brezzi and L. Gastaldi, *Mathematical properties of one-dimensional semiconductors*, Mat. Apl. Comput. **5** (1986), 123–138.

4. F. Brezzi, L. D. Marini, and P. Pietra, *Méthodes d'éléments finis mixtes et méthode de Scharfetter-Gummel*, C. R. Acad. Sci. Paris I **305** (1987), 599–604.

5. ——, *Two-dimensional exponential fitting and applications to semiconductor device equations*, SIAM J. Numer. Anal (to appear).

6. L. Caffarelli and A. Friedman, *A singular perturbation problem for semiconductors*, Boll. Un. Mat. Ital. **1-B** (1987), 409–422.

7. C. Hunt and N. Nassif, *On a variational inequality and its approximation in the theory of semiconductors*, SIAM J. Numer Anal. **12** (1975), 938–950.

8. L. D. Marini, *An inexpensive method for the evaluation of the solution of the lowest-order Raviart-Thomas mixed method*, SIAM J. Numer. Anal. **22** (1985), 493–496.

9. L. D. Marini and A. Savini, *Accurate computation of electric field in reverse-biased semiconductor devices*, Compel **3** (1984), 123–135.

10. P. A. Markowich, *The stationary semiconductor device equations*, Springer, 1986.

11. M. S. Mock, *Analysis of mathematical models of semiconductor devices*, Boole Press, Dublin, 1983.

12. W. V. van Roosbroeck, *Theory of flow of electrons and holes in germanium and other semiconductors*, Bell System Tech. J. **29** (1950), 560–607.

DIP. MECCANICA STRUTTURALE AND I.A.N. DEL C.N.R., PAVIA, ITALY.

Lectures in Applied Mathematics
Volume **25** (1990)

Semiconductor Modelling via the Boltzmann Equation

PIERRE DEGOND, FRÉDÉRIC POUPAUD, BERNARD NICLOT, AND FRÉDÉRIQUE GUYOT

Abstract. In this paper, we give an introduction to the Boltzmann equation of semiconductors, with an emphasis on the collision operator. The connection between the Boltzmann and the drift-diffusion models will be presented in terms of the diffusion approximation, and some mathematical results will be given. Then, we introduce a new numerical method for solving the Boltzmann equation, "the weighted particle method", that may give an interesting alternative to the Monte-Carlo method. Examples of the numerical results will be displayed.

I. Introduction. Most of the numerical simulations of semiconductor devices are based upon the drift-diffusion model (cf. [**1**], [**2**]). However, this model does not fit the experimental data, when applied to submicron or Gallium Arsenide devices (cf. [**3**]). In this paper, we give a presentation of the kinetic model of semiconductors which is expected to give better results in these situations.

The drift-diffusion model is based on Ohm's law, which assumes that the average velocity of the carriers is a function of the electric field ($v(E) = \mu_n(E) \cdot E$, μ_n is the mobility). This relation is obtained at equilibrium, as a consequence of the balance between the free acceleration of the carriers by the electric field, and their scattering by the crystal lattice. However, when the mean path l of the carriers between scatterings is small compared with the active length of the device L, the

1980 *Mathematics Subject Classification* (1985 *Revision*). Primary 65C20; 65M25; Secondary 76P05; 82A40; 82A55.

equilibrium does not have enough time to establish, and Ohm's law is no longer valid.

For instance, in a GaAs device at room temperature, the mean free path is of the order of 10^5 to 10^6 m/s (see **[4]** for numerical values). Thus, for a 1μm long device, the ratio l/L is of the order of 1.

In these conditions, the motion is nearly collisionless (or "ballistic"), and in particular, average velocities v higher than the Ohm's law predicted value $v(E)$ are possible. It is a "velocity overshoot" condition. Reviews of the physical mechanism of these hot electron phenomena can be found in **[5]**, **[6]**, **[7]**.

For hot electron modelling, a modification of the drift-diffusion model has been proposed by many authors in various forms **[6]**, **[8]**, **[9]**. This is the "hydrodynamic model". However, the kinetic model (the Boltzmann equation) seems to give the most accurate description of the physics attainable by numerical computations.

In this paper, we will present the main features of the semiconductor Boltzmann equation, give an outline of its mathematical analysis, and describe a numerical method for its simulation. Some results will be presented. For a more physical description of this model, we refer the reader to **[5]**, **[10]**, **[11]**.

II. The semiconductor Boltzmann equation.

II.1. *The equation.* In this paper, we will only consider electrons, with position x in a bounded open set Ω (the device geometry), and wavevector k belonging to the first Brillouin zone B. We describe the electron population in terms of the number $f(x,k,t)$ of electrons at position x, with wavevector k, at a given time t, averaged over a small volume $dx\,dk$ in the phase space. This average number of electrons is positive, and, by virtue of Pauli's exclusion principle, cannot exceed 1:

$$0 \leq f(x,k,t) \leq 1. \tag{II.1}$$

The evolution of f is governed by the Boltzmann equation

$$\begin{cases} \dfrac{\partial f}{\partial t} + v(k)\cdot\nabla_x f - \dfrac{q}{\hbar}E(x,t)\cdot\nabla_k f = Q(f)(x,k,t), \\ \text{for } x\in\Omega\subset\mathbf{R}^3, \qquad k\in B\subset\mathbf{R}^3, \qquad t\geq 0. \end{cases} \tag{II.2}$$

$v(k)$ is a known vector field giving the velocity associated with a wavevector k. $E(x,t)$ is the electric field, given via Poisson's equation, in terms of the potential $\varphi(x,t)$, and the density $n(x,t)$:

$$\text{(II.3)} \qquad \begin{cases} E = -\nabla_x \Phi, \\ -\Delta\Phi = \dfrac{q}{\varepsilon}(n_d(x) - n(x,t)); \quad \text{for } x \in \Omega, \\ n(x,t) = \displaystyle\int f(x,k,t)(4\pi^3)^{-1}\, dk, \end{cases}$$

and $Q(f)$ is the scattering operator given by

$$\text{(II.4)} \quad Q(f)(x,k,t) = \int_B [s(k',k)f(x,k',t)(1 - f(x,k,t)) \\ - s(k,k')f(x,k,t)(1 - f(x,k',t))]\, dk'.$$

q, $\hbar$, and ε are respectively the electron charge, the reduced Planck constant, and the permittivity of the material. $n_d(x)$ is the doping profile and $s(k,k')$ is a transition rate, which is a known function given by the physical nature of the scattering mechanisms.

We assume that the boundary of Ω can be decomposed into ohmic contacts Ω_d and insulating boundaries Ω_n. On Ω_d, we prescribe the value of the potential and of the incoming part of the distribution function:

$$\text{(II.5)} \qquad \begin{cases} \varphi(x,t) = \varphi_0(x), & x \in \partial\Omega_d, \\ f(x,k,t) = f_0(x,k), & x \in \partial\Omega_d; \quad v(k)\cdot\nu(x) \le 0, \end{cases}$$

where $\nu(x)$ is the outwards unit normal of Ω at x. On $\partial\Omega_n$, we assume reflecting boundary conditions for f and homogeneous Neumann conditions for φ:

$$\text{(II.6)} \qquad \begin{cases} \dfrac{\partial\varphi}{\partial\nu}(x,t) = 0, & x \in \partial\Omega_n, \\ f(x,k,t) = f(x,-k,t), & x \in \partial\Omega_n, \quad k \in B. \end{cases}$$

The conditions on ∂B are periodic. Indeed we recall that the first Brillouin zone is the elementary set of the reciprocal lattice $\widehat{L}$ of the crystal lattice L [**12**]. As a function of k, f must be considered as defined on $\mathbf{R}^3/\widehat{L}$, or equivalently on B, with periodic boundary conditions.

II.2. *Physical meaning of the Boltzmann equation.* Newton's equations of motion for a particle trajectory with no scattering are written:

$$\text{(II.7)} \qquad \frac{dx}{dt} = v(k), \qquad \frac{dk}{dt} = -\frac{q}{\hbar}E(x,t).$$

Thus, equation (II.7) states that the rate of change of $f(x(t),k(t),t)$ is due to scatterings $Q(f)$. Let us review briefly the physical meaning of $v(k)$ and of $Q(f)$.

The velocity $v(k)$ is given by the relation

$$v(k) = \nabla_k \varepsilon(k)/\hbar \tag{II.8}$$

where $\varepsilon(k)$ is the energy of an electron of wavevector k, in the lattice. The $\varepsilon(k)$ relationship is the so-called "band diagram," and is known for most semiconductor materials **[13]**. It is given by the computation of the eigenvalues of a suitable Schrödinger operator **[14]**. In most semiconductors it can be approximated by a parabolic relation (like for the electron in the vacuum)

$$\varepsilon(k) = \frac{\hbar^2 k^2}{2m^*}; \qquad v(k) = \frac{\hbar k}{m^*} \tag{II.9}$$

with an effective mass m^* generally smaller than the electron rest mass m_0 ($m^*/m_0 = 0.98$ in Si and 0.067 in GaAs, cf. **[5]**). In these models it is generally assumed that the Brillouin zone B can be replaced by $\mathbf{R}^3$, with a good accuracy. However, this is not true for all materials, and in particular for GaAs under high electric field conditions. We refer the reader to Section VII of **[15]** and to **[3]** for examples of Boltzmann modellings of multivalley materials.

The collision operator $Q(f)$ describes interactions with defects of the crystal such as ionized impurities, lattice vibrations (or phonons), dislocations,.... (The electron-electron interaction is generally neglected at the usual densities and its precise description would lead us far beyond the scope of this paper. See e.g. **[16]**.) Similar expressions as (II.4) for the collision operator arise in various areas of physics (cf. **[17]**) except for the occurrence of the $(1\text{-}f)$ factors which are consequences of Pauli's exclusion principle. The transition rate $s(k, k')$ is known but is very complicated in the real cases. Indeed, it must account for a large variety of scattering mechanisms, each of them giving rise to a very complicated expression. The precise expressions of $s(k, k')$ can be found in **[5]**. An example is given in Section IV. Let us simply notice that it is generally not a function, but a measure, as a consequence of an energy conservation principle (cf. Section IV).

Finally, let us notice that in most cases, the distribution function f is small compared with 1, so that the Pauli factors can be neglected in the expression of the collision operator (II.4), leading to a simple linear collision operator:

$$Q(f)(x, k, t) = \int [s(k', k) f(x, k', t) - s(k, k') f(x, k, t)]\, dk'. \tag{II.10}$$

In most Monte-Carlo simulations, this form of the collision operator is considered.

II.3. *Main properties of the collision operator.* In this subsection, we describe in more detail the mathematical properties of the collision operator, and their relevance for physical modelling. First, we will investigate the linear form (II.10) and will refer to **[18]** for the proofs. Then, we will give some hints on the mathematical properties of the nonlinear operator (II.4). Throughout this section, we will assume that $s(k, k')$ is regular and satisfies

$$s(k', k) = s(k, k') \exp(\varepsilon(k') - \varepsilon(k)). \tag{II.11}$$

For simplicity we will scale the physical constants to 1, and assume a parabolic band structure (II.9) with $k \in \mathbf{R}^3$. Although in the physical cases, $s(k, k')$ is not regular, it still satisfies (II.11) and we expect the results presented below to still be relevant.

Introducing the Maxwellian

$$M(k) = (2\pi)^{-3/2} \exp(-k^2/2) \tag{II.12}$$

we can use relation (II.11) to write the collision operator (II.10) in the form

$$Q(f)(k) = \int \varphi(k, k')(M(k)f(k') - M(k')f(k))\, dk' \tag{II.13}$$

where $\varphi(k, k')$ is a regular symmetric function. Note that Q operates on the k variable, and will thus be viewed as acting on a space of functions of k only. Wc introducc thc scattcring frcqucncy $\lambda(k)$ by

$$\lambda(k) = \int s(k, k')\, dk' = \int \varphi(k, k')M(k')\, dk' \tag{II.14}$$

and we define the following spaces

$$\begin{cases} X = \{f; \lambda^{1/2} M^{-1/2} f \in L^2(\mathbf{R}^3)\}, \\ Y = \{g; \lambda^{-1/2} M^{-1/2} g \in L^2(\mathbf{R}^3)\}. \end{cases} \tag{II.15}$$

Then we have:

PROPOSITION 1.

(i) *Q maps continuously X into Y and satisfies for any f_1 and f_2 in X*:

$$\int Q(f_1) f_2 M^{-1}\, dk = \int Q(f_2) f_1 M^{-1}\, dk. \tag{II.16}$$

(ii) *The kernel of Q is spanned by $M(k)$ and its range is defined by*

$$R(Q) = \left\{ g \in Y, \int g(v)\, dv = 0 \right\}. \tag{II.17}$$

(iii) $-Q$ *is coercive on* $(\ker Q)^{\perp}$, *that is, we have* (*with* $\sigma > 0$)

$$-\int Q(f) f M^{-1}\, dk \geq \sigma \int f^2 \lambda M^{-1}\, dk \tag{II.18}$$

for any function f *in* X *such that*

$$\int \lambda f\, dk = 0. \tag{II.19}$$

In physical terms, the Maxwellian $M(k)$ is the equilibrium solution ($Q(M) = 0$, prop. (ii)) and the collision operator preserves mass (II.17). Thus Q cannot be selfadjoint in L^2, but is selfadjoint as an operator from X to Y (II.16), and this expresses some symmetry between the $k \to k'$, and $k' \to k$ processes, resulting from hypothesis (II.11). Finally, (II.18) states that any space homogeneous disturbance of the equilibrium $M(k)$ will be damped at rate $\exp(-\sigma t)$; this is the relaxation process towards equilibrium.

From Proposition 1, it is easy to prove by semigroup theory that equation (II.2) with given electric field E admits a global positive solution satisfying L^1 and L^∞ estimates **[18]**.

Now, we turn to the nonlinear operator (II.4) which can be written in a similar form as (II.13), according to:

$$Q(f)(k) = \int \varphi(k,k')[M(k)f(k')(1-f(k)) - M(k')f(k)(1-f(k'))]\, dk' \tag{II.20}$$

By a simple iterative argument, it is easy to prove that (II.1) will hold for any time, provided it holds for the initial data (this is not true for the linearization (II.13)). From this, and in the case of a bounded Brillouin zone, one can derive a global existence result for the full set of Boltzmann–Poisson equation (II.2) to (II.4) (cf. **[19]**). Again, (II.20) preserves mass:

$$\int Q(f)(k)\, dk = 0$$

and its kernel is now a nonlinear manifold, consisting of Fermi–Dirac distribution functions:

$$\operatorname{Ker} Q = \left\{ \left(1 + \exp\left(\frac{k^2}{2} - \mu\right)\right)^{-1},\ \mu \in \mathbf{R} \right\}.$$

The parameter μ is the so-called Fermi energy or chemical potential, and is of course related to the density.

Most of the properties described above are standard in kinetic theory, and analogues of them in other areas can be found for instance in **[17]**, **[20]**.

III. The diffusion approximation.

III.1. *Introduction.* The drift-diffusion model is expected to be a good approximation of the Boltzmann equation, when the mean free path of the carrier is small. The purpose of this section is to outline a mathematically rigorous treatment of this approximation problem. The method is based on the diffusion approximation, which has been used by physicists for a long time in various areas [**17**]; for its application to semiconductors, we refer to [**10**], [**11**], [**21**]. The mathematical theory has been set by Larsen–Keller [**22**], Bensoussan–Lions–Papanicolaou [**23**] and Bardos–Santos–Sentis [**24**]. Its application to semiconductors is detailed in [**18**].

Let us point out that an intermediate model between the Boltzmann equation and the drift-diffusion equation is often used and referred to as the "hydrodynamical model" or the "displaced Maxwellian theory", [**6**], [**8**], [**9**]. It essentially incorporates more moments of the distribution function than the drift-diffusion model. However, no mathematical theory is yet available to justify this approximation.

III.2. *The diffusion approximation.* Introducing the mean time between collisions τ, we expect the drift-diffusion model to be valid when τ is small (i.e., the collisions are frequent) and when we investigate large time scales compared with τ. In these conditions, it is natural [**22**] to scale the equation (in dimensionless units) according to

$$\frac{\partial f^\tau}{\partial t} + \frac{1}{\tau}(k \cdot \nabla_x f^\tau - E \cdot \nabla_k f^\tau) = \frac{1}{\tau^2} Q(f^\tau) \tag{III.1}$$

and to expand f^τ into powers of τ:

$$f^\tau = f_0 + \tau f_1 + \tau^2 f_2 + r_\tau . \tag{III.2}$$

The functions $f_i(x,k,t)$ for $i = 0, 1, 2$ are determined by plugging (III.2) into (III.1) and identifying coefficients of equal powers of τ. This gives

$$Q(f_0) = 0, \tag{III.3}$$

$$Q(f_1) = k \cdot \nabla_x f_0 - E \cdot \nabla_k f_0, \tag{III.4}$$

$$Q(f_2) = k\nabla_x f_1 - E \cdot \nabla_k f_1 + \frac{\partial f_0}{\partial t}. \tag{III.5}$$

Equation (III.3) is solved thanks to Proposition 1-(ii) and leads to

$$f_0(x,k,t) = n(x,t)M(k)$$

where $M(k)$ is the Maxwellian (II.12) and $n(x,t)$ is a function to be determined.

Equation (III.4) now reads

$$Q(f_1) = (\nabla_x n - nE) \cdot k \cdot M(k) \tag{III.6}$$

and since $(\nabla_x n - nE)$ is independent of k, (III.6) can be solved by

$$f_1 = (\nabla_x n - nE) \cdot h(k) \tag{III.7}$$

where $h(k) = (h_i(k))_{i=1}^3$ is a solution of

$$Q(h_i) = k_i M(k), \qquad i = 1, 2, 3. \tag{III.8}$$

This equation is solvable because $kM(k)$ matches the solvability condition (II.17).

Finally, equation (III.5) can be solved if the right-hand side matches the solvability condition (II.17), that is, if

$$\int \left(k \cdot \nabla_x f_1 - E \nabla_k f_1 + \frac{\partial f_0}{\partial t} \right) dk = 0. \tag{III.9}$$

The middle term vanishes after integration. By defining a diffusivity tensor

$$D_{ij} = -\int k_i h_j(k)\, dk, \tag{III.10}$$

we can write (III.9) as

$$\frac{\partial n}{\partial t} - \nabla_x D(\nabla_x n - nE) = 0, \tag{III.11}$$

which is precisely the drift-diffusion equation.

The physical interpretation of this expansion is the following. In the limit of zero mean free path, the zeroth-order term is a local equilibrium (III.6), defined by the Maxwellian at the lattice temperature $M(k)$, and by a possible space inhomogeneous time dependent density $n(x, t)$. The first-order term is related to the current density (proportional to $\nabla_x n - nE$). Now, this theory derives the drift-diffusion equation from a solvability condition for the second-order term f_2.

Now, it is possible to estimate the remainder r_τ and to show that it is small. More precisely we have

PROPOSITION 2 [**18**]. *Under the hypotheses of Section* II.3, *and if the electric field is a smooth function of* x, *we denote by* f^τ *the solution for* t *in* $[0, T]$ *of equation* (III.1) *with initial and boundary conditions given by*

$$\begin{cases} f^\tau(x, k, 0) = f_0(x) \cdot M(k); & x \in \Omega, \\ f^\tau(x, k, t) = g(x, t) \cdot M(k); & x \in \partial\Omega, \quad k \cdot \nu(x) < 0, \end{cases} \tag{III.12}$$

and satisfying

$$g \in L^\infty(\partial\Omega \times [0,T]); \qquad f_0 \in L^\infty(\Omega). \tag{III.13}$$

We denote by $n(x,t)$ the solution of equation (III.11) *with initial and boundary conditions*

$$\begin{cases} n(x,0) = f_0(x); & x \in \Omega, \\ n(x,t) = g(x,t); & x \in \partial\Omega. \end{cases} \tag{III.14}$$

Then there exists a constant independent of τ such that

$$\|f^\tau(x,k,t) - n(x,t)M(k)\| \le C\tau$$

in the space $L^1([0,T] \times \Omega \times \mathbf{R}^3)$.

Of course, the diffusivity tensor (III.10) is positive definite and if s is invariant under rotations, reduces to a scalar. Furthermore in equation (III.11), the mobility appears to be equal to the diffusivity. Going back to the usual units this gives the Einstein relation.

III.3. *Boundary layers.* Conditions (III.12) may appear quite restrictive, since they require the initial-boundary data to be local thermodynamical equilibrium (i.e., proportional to $M(k)$ as functions of k). However, investigating general initial-boundary conditions only adds initial-boundary layers of thickness τ, in which the collision operator relaxes strongly the solution towards a local equilibrium. In order to make this concept clearer we have studied boundary layers. We consider a neighbourhood of a point ξ on the boundary $\partial\Omega$, with unit outward normal $\nu(\xi)$. If we expand the space scale along the axis spanned by $\nu(\xi)$, in order to make the structure of the boundary layer clearer, we are led at leading order to the problem

$$\begin{cases} k_1 \dfrac{\partial f}{\partial x} = Q(f); & x \in [0,\infty[, \qquad k \in \mathbf{R}^3 \\ f(0,k) = g_0(k); & k \in \mathbf{R}^3 \quad \text{such that } k_1 > 0 \end{cases} \tag{III.15}$$

where x is the position along the axis spanned by $\nu(\xi)$, and $x = 0$ is the boundary point. The incoming flux $g_0(k)$ is thus defined for $k_1 \geq 0$, where k_1 is the projection of k onto the x-axis. Problem (III.15) is often referred to as the Milne problem, especially in the context of radiative transfer.

It is convenient to introduce the decomposition of $f(x,k)$ on $\operatorname{Ker} Q$ and $(\operatorname{Ker} Q)^{\perp}$, according to the inner product of X:

$$\text{(III.16)} \qquad \begin{cases} f(x,k) = \alpha(x)M(k) + g(x,k), \\ \displaystyle\int \lambda(k)g(x,k)\,dk = 0, \\ \alpha(x) = \displaystyle\int \lambda(k)f(x,k)\,dk \Big/ \int \lambda(k)M(k)\,dk. \end{cases}$$

In addition to the hypotheses of Proposition 1, we assume the following:

$$\text{(III.17)} \qquad c_1 \geq \lambda(k) \geq c_2/(1+|k|)^m, \qquad m > 0,$$

$$\text{(III.18)} \qquad |g_0(k)| \leq c|k|^n M(k), \qquad n > 0.$$

Then we have

PROPOSITION 3 **[18]** (Existence and uniqueness). *There exists a unique function f of problem* (III.15) *such that its decomposition* (α, g) *satisfies*

$$\text{(III.19)} \quad g \in L^2([0,\infty[, X); \qquad \int_0^{\infty} \int |g(x,k)|^2 \lambda M^{-1}(k)\,dk \leq C(g_0)$$

and such that there exists $\alpha_\infty \in \mathbf{R}$ satisfying

$$\text{(III.20)} \qquad \alpha(x) - \alpha_\infty \in L^2([0,\infty[).$$

PROPOSITION 4 **[18]** (Asymptotic decay). *For the solution f defined at Proposition* 3, *we have*

$$\text{(III.21)} \qquad |g(x,\cdot)|_X + |\alpha(x) - \alpha_\infty| \leq C(g_0)\exp(-\gamma x^\beta)$$

where $\beta = 2/(m+3)$ and γ is a positive constant. We denote

$$\text{(III.22)} \qquad |g(x,\cdot)|_X = \left(\int |g(x,k)|^2 \lambda M^{-1}(k)\,dk\right)^{1/2}.$$

Proposition 4 states that the solution of (III.15) is the superposition of a global equilibrium state $\alpha_\infty M(k)$ and of a fast varying solution given by $g(x,k) + (\alpha(x) - \alpha_\infty)M(k)$, decaying at the almost exponential rate $\exp(-\gamma x^\beta)$. Only a subexponential rate ($\beta < 1$) is obtained because of the decay of the scattering frequency at high velocities (III.17). Thus, it takes a longer distance to the collision operator to relax the high velocities than the low velocities. This explains the dependence of β upon the decay rate m of λ.

Proposition 4 explains (a posteriori) why the existence and uniqueness result is not obvious. Indeed from (III.21) we see that the full solution f is not in $L^2([0,\infty[\ ,X)$, and thus, standard energy estimates are not available for problem (III.15). Uniqueness is not obvious either. Indeed the solutions found in Proposition 3 have vanishing fluxes, $\int k_1 f\,dk = 0$. However, other solutions with the same boundary data $g_0(k)$ but with nonvanishing fluxes can be obtained.

As a conclusion, we notice that such a theory can be used to design proper boundary conditions for the drift-diffusion equation, which take into account the effects of the boundary layer more accurately than (III.14). (cf. e.g. [**24**]). This could be particularly useful for the interface conditions at a heterojunction.

IV. Numerical simulations of the Boltzmann equation.

IV.1. *Introduction.* The most widely used numerical method for the Boltzmann equation (II.2) is the Monte-Carlo method. Introductory papers to this method can be found in [**5**], and examples of its application in [**4**], [**5**], [**25**], [**26**]. However, the need for other methods is rising for several reasons: the random choices introduce a lot of numerical noise; extensions of the method to nonlinear situations, such as the Pauli-degenerate operator (II.4) are very difficult; the resolution of sharp gradients of the doping density requires a tremendous amount of particles.

From our knowledge, only a few alternate methods have been investigated [**7**], [**27**], [**38**] and some of them can only be used in very particular situations.

Our aim in this section is to introduce a new method, the "weighted particle method", designed by S. Mas-Gallic [**28**] following the ideas of P. A. Raviart et al [**29**], [**30**].

IV.2. *The weighted particle method* [**28**]. An elementary introduction to particle methods may be found in [**31**]. Let $f_h(x,k,t)$ be the approximation of f into Dirac masses (particles):

$$f_h(x,k,t) = \sum_i (\omega_i f_i(t))\delta(x - x_i(t))\delta(k - k_i(t)), \tag{IV.1}$$

where for each particle i, we denote by

$(x_i(t), k_i(t))$	its position in the phase space (x,k),
ω_i	its control volume (constant in t),
$f_i(t)$	its weight.

ω_i is, roughly speaking, the volume of the phase space that the particle carries, and $f_i(t)$ is the number of physical particles that are represented by the numerical particle.

The positions $(x_i(t), k_i(t))$ follow the characteristics of the first-order differential part of equation (II.2):

$$\text{(IV.2)} \qquad \begin{cases} \dfrac{dx_i}{dt} = v(k_i); & x_i(0) = x_i^0, \\ \dfrac{dk_i}{dt} = E(x_i, t); & k_i(0) = k_i^0, \end{cases}$$

which implies that

$$\text{(IV.3)} \qquad \left(\frac{\partial}{\partial t} + v(k)\cdot\nabla_x - \frac{q}{\hbar}E(x,t)\cdot\nabla_k\right)\cdot f_h = \sum_i \omega_i \frac{df_i}{dt}\delta(x - x_i(t))\delta(k - k_i(t)).$$

Thus, denoting by Q_i an approximation of the collision operator $Q(f)(x_i, k_i, t)$, we deduce from (IV.3) and (II.2) that the weights must evolve according to

$$\text{(IV.4)} \qquad \frac{df_i}{dt} = Q_i.$$

An estimate of Q_i can be found by replacing the integral in (II.4) by a numerical quadrature where the particles are the quadrature points

$$\text{(IV.5)} \quad Q_i = \sum_j [S(k_j, k_i) f_j (1 - f_i) - S(k_i, k_j) f_i (1 - f_j)] \zeta_\varepsilon (x_j - x_i)\omega_j.$$

Here, ζ_ε is a smooth approximation of the Dirac measure,

$$\text{(IV.6)} \qquad \zeta_\varepsilon(x) \to \delta(x) \quad \text{as} \quad \varepsilon \to 0 \quad \text{in } \mathscr{D}'(\mathbf{R}^3),$$

which takes into account the fact that the collision operator is integral in k but not in x. (No integration over dx is performed in II.4.) A possible choice for ζ_ε is to take

$$\text{(IV.7)} \qquad \zeta_\varepsilon(x) = \varepsilon^{-3}\zeta(x/\varepsilon); \qquad \int_{-1}^{1} \zeta(x)\,dx = 1$$

where $\zeta(x)$ is a smooth localized function (for instance a B-spline).

Thus, the weighted particle method reads

$$\text{(IV.8)} \qquad \begin{cases} \dfrac{dx_i}{dt} = v(k_i), \\ \dfrac{dk_i}{dt} = E(x_i, t), \\ \dfrac{df_i}{dt} = \displaystyle\sum_j [S(k_j, k_i) f_j (1 - f_i) \\ \qquad\qquad - S(k_i, k_j) f_i (1 - f_j)] \zeta_\varepsilon (x_j - x_i) \omega_j. \end{cases}$$

The system of ordinary differential equations (IV.8) is solved by classical methods (Runge–Kutta or multistep methods). The electric field $E(x_i, t)$ has generally to be computed self-consistently with the position of the particles via Poisson's equation (II.3). For that purpose the classical particle in cell (P.I.C.) method is generally used. It consists of using a mesh to solve Poisson's equation, and assignment-interpolation procedures to allow the information to circulate from the particles to the mesh and vice versa (cf. **[32]**, **[33]**, **[34]** for more details). An alternate method has been proposed, using Green functions **[35]**. The numerical analysis method (IV.8) has been performed in **[28]**, and preliminary tests on the neutron-transport equation can be found in **[36]**.

IV.3. *A test case. The homogeneous field model.* The first application of the weighted-particle method for the semiconductor Boltzmann equation deals with the homogeneous field model **[15]**. It describes a one-dimensional homogeneous and infinite sample of semiconductor submitted to a constant external electric field E. Thus, no space variable is introduced and no Poisson equation has to be solved. Furthermore, we assume that f is invariant under rotations of axis E; thus, we can use cylindrical coordinates, and consider only two momentum coordinates $(k_\parallel, k_\perp)$: the longitudinal and transverse components of k with respect to E. This leads to the following model:

$$\begin{cases} \dfrac{\partial f}{\partial t} - \dfrac{qE}{\hbar} \dfrac{\partial f}{\partial k_\parallel} = Q(f), \\ Q(f) = \int [\sigma(k', k) f(k')(1 - f(k)) \\ \qquad\qquad - \sigma(k, k') f(k)(1 - f(k'))] 2\pi k_\perp \, dk_\parallel \, dk_\perp, \end{cases}$$

where k stands now for $(k_\parallel, k_\perp)$ and σ is a reduced transition rate obtained by averaging s over the angular variable **[15]**.

Practical considerations on the application of the weighted-particle method to (IV.9) are given in **[15]**. Also some comparisons with analytical solutions which can be obtained for particularly simple σ are displayed in [15]. As an example, we wish to present here a real model of scattering, and the results of its simulation by the weighted-particle method. We do not pretend at all that the present results are relevant for real devices. We only want to show that reasonable results can be expected from this method.

IV.4. *The polar optical scattering.* This model describes the interaction of electrons and optical phonons. The transition rate is given by [5]:

$$\text{(IV.10)} \quad S(k,k') = \frac{S_{\text{opt}}}{|k-k'|^2}[(N_0+1)\delta(\varepsilon(k')-\varepsilon(k)+\varepsilon_0) + N_0\delta(\varepsilon(k')-\varepsilon(k)-\varepsilon_0)]$$

where S_{opt}, N_0, and ε_0 are constants. The Dirac masses express the conservation of total energy of the system {electron, phonon} in the course of the interaction. Thus s is not a function but a measure. As a consequence of this singularity, the solution has very few regularity properties [37] and the numerical approximation must involve an additional smoothing,

$$\text{(IV.11)} \quad S(k,k') \simeq \frac{S_{\text{opt}}}{|k-k'|^2}[(N_0+1)\zeta_\alpha(\varepsilon(k')-\varepsilon(k)+\varepsilon_0) + N_0\zeta_\alpha(\varepsilon(k')-\varepsilon(k)-\varepsilon_0)],$$

where ζ_α is an approximation of the Dirac mass, like in (III.6). The influence of this additional smoothing is investigated theoretically in [37], and numerically in [15]. Only a linear collision operator (II.10) was considered in this simulation.

The initial data was chosen to be Maxwellian (cf. (II.12)). Thus, this simulation models the transient response of a sample of semiconductor submitted to a step of electric field:

$$\text{(IV.12)} \qquad E(t) = 0 \quad \text{for } -\infty \le t \le 0; \qquad E(t) = E \text{ for } t > 0.$$

Figure 1 displays the mean velocity $\bar{v}$ and mean energy $\bar{\varepsilon}$ of the solution versus time. $\bar{v}$ and $\bar{\varepsilon}$ are defined according to

$$\text{(IV.13)} \qquad \begin{cases} \bar{v}(t) = \dfrac{\hbar}{m^*}\displaystyle\int k_\parallel f(k,t) 2\pi k_\perp \, dk, \\ \bar{\varepsilon}(t) = \dfrac{\hbar^2}{2m^*}\displaystyle\int k^2 f(k,t) 2\pi k_\perp \, dk. \end{cases}$$

The numerical values were those of a moderately doped GaAs sample at 77K (cf. [15]). These curves present a slight velocity and energy overshoot at time 1.5 ps which can be explained by analyzing the evolution

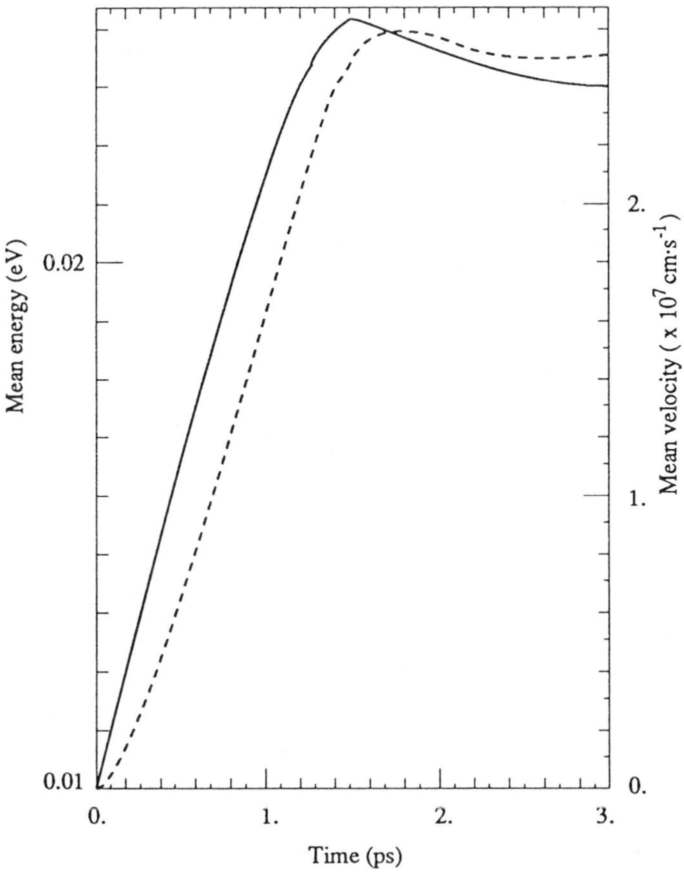

FIGURE 1. Mean velocity $\bar{v}$ (solid line) and mean energy $\bar{\varepsilon}$ (dashed line) versus time for the polar optical interaction (IV.10). The value of the phonon energy ε_0 is 0.036 eV, and the electric field $E = 10^5$ Vm^{-1}. 1200 particles were used and the value of the smoothing parameter α in (IV.11) chosen to be $0.025 \times \varepsilon_0$. Slight velocity and energy overshoots are present, at time $t = 1.5$ ps, due to the sudden turn-on of the interaction when the bulk of the electrons reaches the energy ε_0.

of the distribution function (cf. Figure 2). Indeed the scattering rate $\lambda(k)$ for the interaction (IV.11) is much higher for $\varepsilon(k) > \varepsilon_0$ than for $\varepsilon(k) < \varepsilon_0$. Thus, as long as the peak of the initial Maxwellian has not reached this threshold enery ε_0, almost no scattering occurs, and the distribution function is almost a displaced Maxwellian (Fig. 2.1). As soon as this threshold energy has been reached, strong optical scattering begins to occur, which scatters back the electrons towards the almost zero energies (Fig. 2.2). This reduces the average velocity and energy of the carriers and explains the velocity overshoot. In Fig. 2.3 the final stationary state has been reached, and the distribution function exhibits a sharp gradient near the energy ε_0.

IV.5. *Some nonlinear cases.* Still dealing with the homogeneous field model, we present here the results of the weighted particle method in two nonlinear cases. The collision operator is the Pauli degenerate operator (II.4) and, in both cases, the initial data is a Fermi–Dirac distribution function

$$f_0 = (1 + \exp((\varepsilon(k) - \mu)/k_b T))^{-1}, \tag{IV.14}$$

where μ is the chemical potential, related to the density. The relaxation time model is the simplest model that one can think of. It is simply given by $\varphi(k, k')$ (cf. formula (II.13)) constant:

$$\varphi(k, k') = \lambda. \tag{IV.15}$$

The nonlinear collision operator (II.4) is given in that case by

(IV.16)

$$\begin{aligned} Q(f)(k) &= \int \lambda[M(k)f(k')(1 - f(k)) - M(k')f(k)(1 - f(k'))]\, dk' \\ &= -[f(1 - \alpha + nM) - nM] \end{aligned}$$

where

$$\alpha = \frac{1}{4\pi^3} \int f(k)M(k)\, dk$$

and n is the density (cf. (II.3)).

The corresponding linear operator (II.10) is given by

$$\begin{aligned} Q(f)(k) &= \int \lambda[M(k)f(k') - M(k')f(k)]\, dk' \\ &= -\lambda(f - nM). \end{aligned} \tag{IV.17}$$

The effect of the degeneracy factors $(1 - f)$ is to reduce the transition rate. Indeed, replacing $s(k', k)f(k')(1 - f(k))$ by $s(k', k)f(k')$ in the linearization means overestimating the transition rate from k' to k and therefore the collision effects. Figure 3 displays the mean velocity $\bar{v}$ versus time, in the linear and nonlinear cases. In the nonlinear case, the stationary mean velocity is higher and the relaxation towards the stationary state is slower. In particular, we expect that the stationary velocity $\bar{v}_s$ is now dependent upon the doping density (which is untrue in the linear case). To highlight this dependence, we provide a few lines of

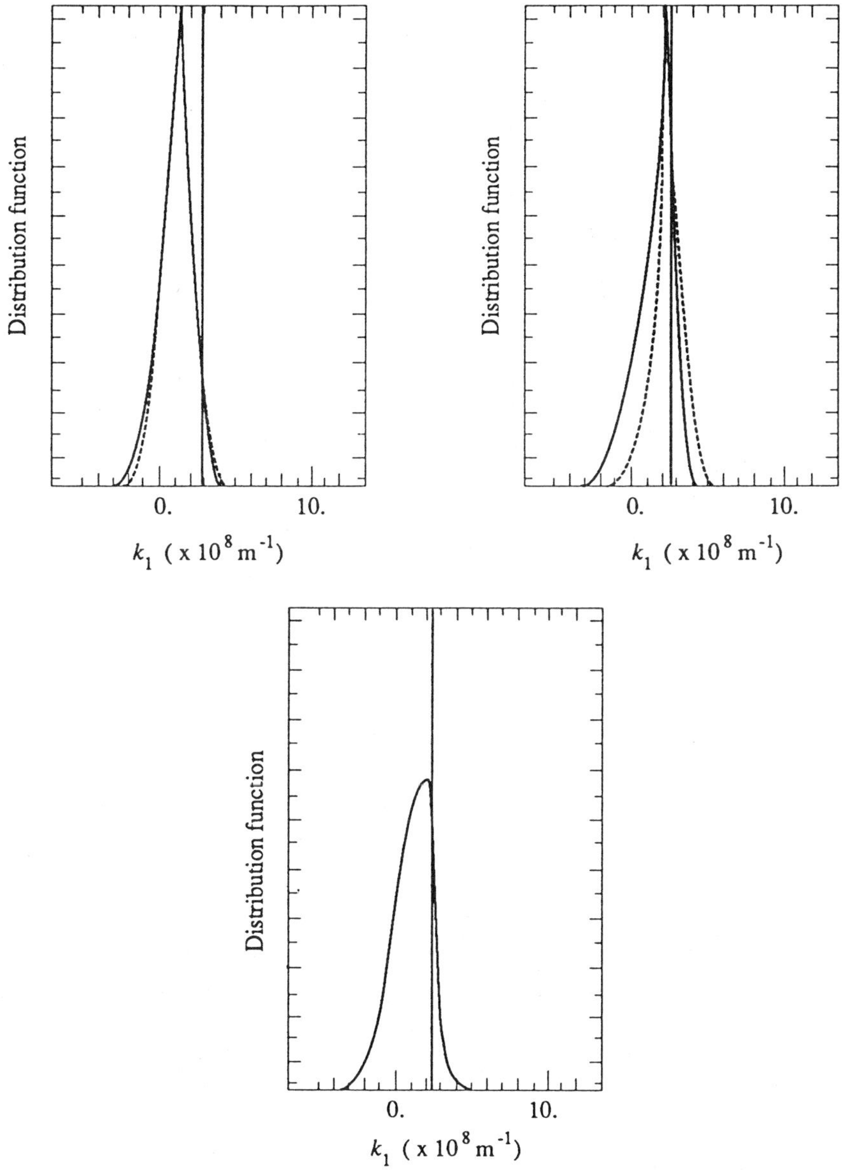

FIGURE 2. Trace of the distribution function along the k_1 axis (parallel to the electric field) plotted in arbitrary units: on Fig. 2.1 at time 1 ps; on Fig. 2.2 at time 1.6 ps; on Fig. 2.3 at time 3 ps (stationary state). With the dashed line is plotted a displaced Maxwellian with the same location of the peak. The vertical line represents the value of the threshold energy ε_0. The numerical parameters are the same as in Figure 1.

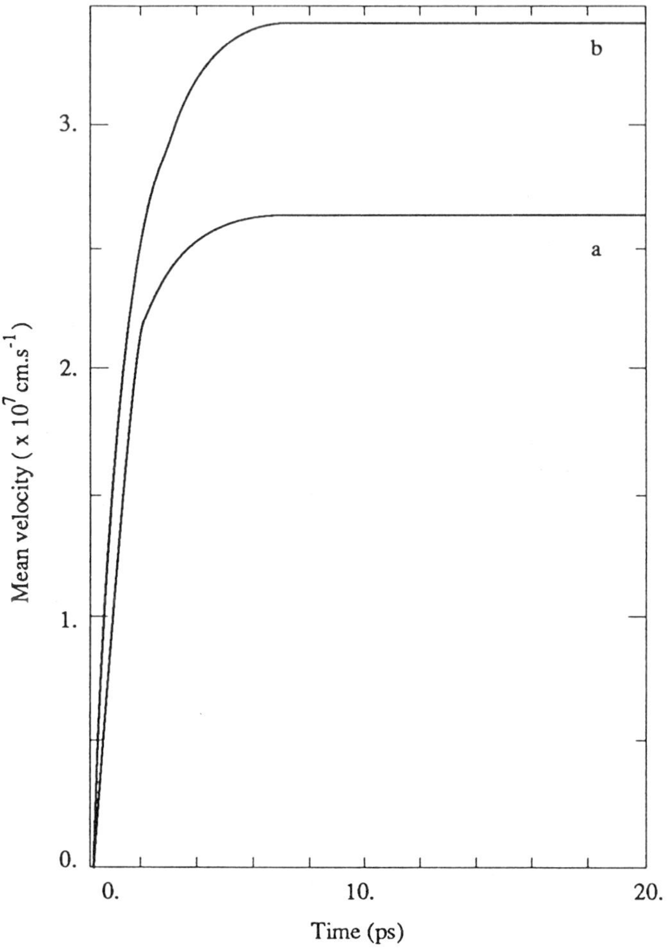

FIGURE 3. Mean velocity versus time for the relaxation time model, in the linear approximation (IV.17): curve a, and in the nonlinear case (IV.16): curve b. $\lambda = 10^{12}\ s^{-1}$; $E = 10^5\ Vm^{-1}$. 2000 particles were used. In the nonlinear case b, the relaxation is slower and the stationary velocity is higher.

analysis which give the asymptotic behaviour of $\bar{v}_s$ for very high doping densities and low electric fields. This analysis relies on the diffusion approximation which has been extensively presented in Section III.

We consider the stationary state equation after scaling:

$$\tau \frac{\partial f}{\partial k_{\|}} = f(1 - \alpha + nM) - nM \qquad \text{(IV.18)}$$

and τ is actually the ratio of the drift velocity v_d to the thermal velocity v_{th}

$$\tau = v_d/v_{th} = \frac{qE}{\lambda(K_b T m^*)^{1/2}}. \tag{IV.19}$$

Performing the expansion as in Section III.2, we find that

$$f = f_0 + \tau f_1$$

where f_0 is a Fermi–Dirac distribution function (cf. (IV.14)) and f_1 can be chosen the following way:

$$f_1 = f_0 \frac{\partial f_0}{\partial k_1} \Big/ (n_0 M) \tag{IV.20}$$

where

$$n_0 = \frac{1}{4\pi^3} \int f_0(k) 2\pi k_\perp \, dk_\parallel \, dk_\perp.$$

Since f_0 is an even function of $k_\parallel$ and f_1 is an odd function of $k_\parallel$, from those expressions we find that

$$\bar{v}_s = \tau \int f_1 2\pi k_\perp \, dk_\parallel \, dk_\perp, \tag{IV.21}$$

$$n = \int f_0 2\pi k_\perp \, dk_\parallel \, dk_\perp.$$

For high values of the chemical potential μ (and therefore of the doping density n) we deduce from formulae (IV.14) and (III.20) the following asymptotic behaviour (after some tedious calculations):

$$\bar{v}_s \sim B\mu^{-3/2} \exp(\mu/(K_b T)),$$
$$n \sim A\mu^{3/2},$$

where A and B are related to the electric field and to the temperature. From these relations we expect the dependence of

$$(\operatorname{Log} n\bar{v}_s - \operatorname{Log} C)^{3/2} \tag{IV.22}$$

versus n to be linear, where

$$C = (K_b T/(2\pi))^{3/2} qE(m^*)^{1/2}/(\lambda \hbar^3).$$

The numerical simulation has been performed with an electric field of 10^2V/m and a constant scattering frequency $\lambda = 10^{12}\text{s}^{-1}$. The physical parameters were those of Gallium-Arsenide at a lattice temperature $T =$ 77K. In order to check the linear asymptotic dependence of the quantity (IV.22) versus n, we choose fairly high values of the chemical potential μ, from $7 \times K_b T$ to $9 \times K_b T$, which leads to doping densities ranging

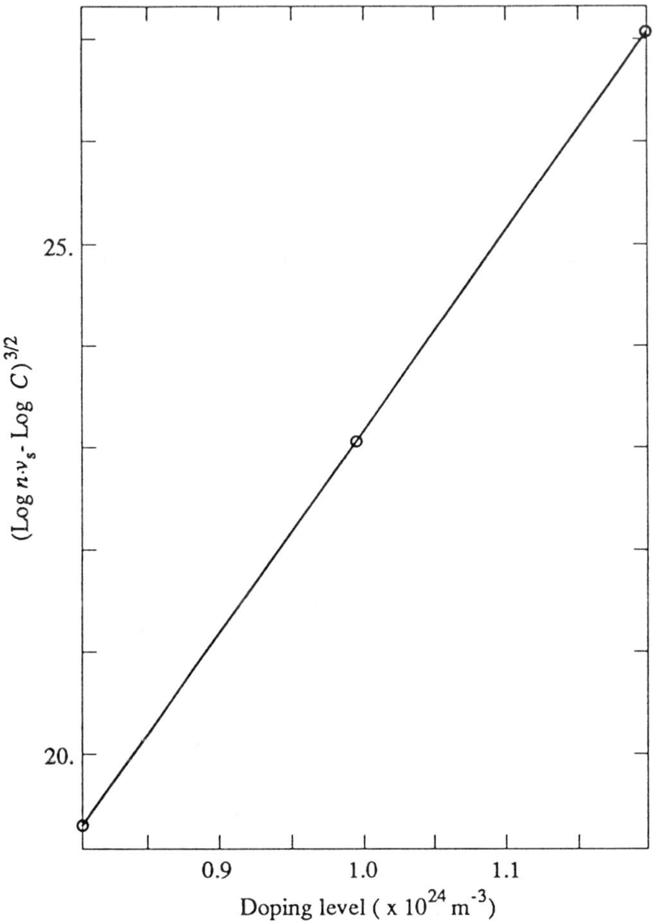

FIGURE 4. $(\text{Log}\, nv_s - \text{Log}\, C)^{3/2}$ versus doping level n for the nonlinear relaxation time model (IV.16). The diffusion approximation, and an asymptotic analysis predicts that this relation should be linear, for small electric fields and very high densities. Numerical computations performed with $E = 10^2$ Vm^{-1} and three values of the density (indicated by circles on the figure) confirm this prediction. The departure from linearity is only 2%.

from 10^{23}m^{-3} to 10^{24}m^{-3}. The result of the computation is shown in Figure 4 and gives a fairly good agreement with the prediction (IV.22). The relative error between the computed values of the quantity (IV.22) and those obtained performing a linear extrapolation is less than 2%.

We now present the nonlinear case of polar optical scattering. The collision operator is that given in (II.4) and the transition rate is the same as (IV.10). Figure 5 displays the mean velocity $\bar{v}$ versus time for different doping levels: $10^{22}/m^3$, $10^{24}/m^3$, and $3 \cdot 10^{24}/m^3$.

When the doping level goes from $10^{22}/m^3$ to $10^{24}/m^3$, the stationary mean velocity decreases. We can explain this the following way: the chemical potential which is at first much lower than the optical threshold increases and comes near the threshold; moreover, the distribution function becomes wider and thus a larger part of it is above the threshold. Therefore more electrons undergo the optical scattering earlier, so that both the stationary velocity and the overshoot magnitude decrease. But when the doping level goes still higher, the chemical potential is much larger than the threshold; thus the threshold influence vanishes. In this regime the explanations that we gave for the relaxation time model become relevant and the stationary velocity increases with the doping level.

These examples were intended to show that direct simulations of the Boltzmann equation could provide an insight into the physics which could be of great help for future device modelling. In particular, the nonlinear equation obtained with the Pauli degenerate operator does not bring out major problems when using the weighted-particle method. Therefore the latter may be actually useful for simulations of highly doped components. Another example of intervalley scattering can be found in [**15**].

V. *Conclusion.* In this paper, we presented an introduction and some mathematical analysis of the semiconductor Boltzmann equation, and we outlined a new numerical method to solve the Boltzmann equation. We think that semiconductor modelling via the Boltzmann equation will become more and more important in the next few years, especially for the analysis of the new structures that are now technologically available: heterojunctions, superlattices, two-dimensional electron gas devices,.... By no means is the drift-diffusion model relevant for these structures and we expect the Boltzmann equation to play the same role for these structures as the drift-diffusion model did for silicon devices. It is therefore important to perform a mathematical analysis, and to provide accurate and practically feasible numerical methods for the Boltzmann equation.

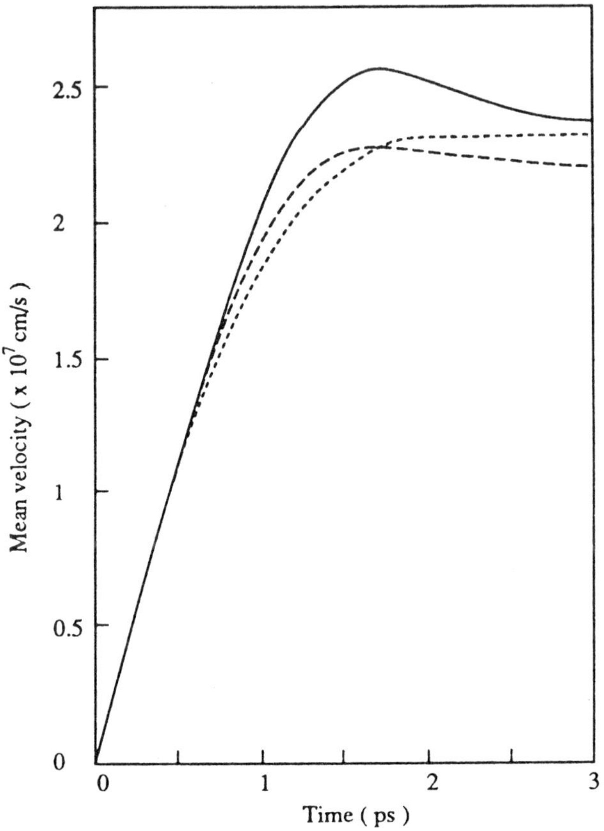

FIGURE 5. Mean velocity versus time for the polar optical scattering (IV.10), in the nonlinear case (II.4). The same physical constants were used as in the linear case (displayed on Figure 1). The computation was performed with 960 particles, $\alpha = 0.1 \times \varepsilon_0$, and for three values of the electron density. For a density of $5 \cdot 10^{22}\mathrm{m}^{-3}$ (solid line), the mean velocity exhibits the same overshoot behaviour as in the linear case. At $10^{24}\mathrm{m}^{-3}$ (dashed line), the overshoot has disappeared and the stationary velocity is lower. However, in the case $6 \cdot 10^{24}\mathrm{m}^{-3}$ (dotted line), the stationary mean velocity increases again and the behaviour of the transient response is very near to that of a constant relaxation time model. The nonmonotonous behaviour of the stationary velocity when the doping density is increased can be explained by examining the relative position of the Fermi energy with respect to the optical phonon energy (see text).

REFERENCES

1. P. A. Markowich, *The stationary semiconductor device equations*, Springer-Verlag, Wien, New York, 1986, and S. Selberherr, *Analysis and simulation of semiconductor devices*, Springer-Verlag, Wien, New York, 1984.

2. *IEEE Trans. on Electron Devices*, ED-30 (1983);ED-32 (1985).

3. J. L. Blue and C. L. Wilson, *Modelling Gallium-Arsenide Transistors*, this volume.

4. P. Hesto, *Injection dependence of quasiballistic transport in GaAs at* 77K, Surf. Science **132** (1983), 623.

5. L. Reggiani (ed.), *Hot-electron transport in semiconductors*, Topics in Applied Physics Series, Springer, Berlin, 1985.

6. M. S. Shur and L. F. Eastman, *Ballistic transport in semiconductor at low temperature for low-power high-speed logic*, IEEE Trans. on Electron Devices **ED-26** (1979), 1677; and *Near ballistic electron transport in GaAs devices at* 77K, Solid State Electron **24** (1981), 11.

7. H. U. Baranger and J. W. Wilkins, *Ballistic electrons in an inhomogeneous submicron structure: Thermal and contact effects*, Phys. Rev. **B-30** (1984), 7349; and *Ballistic structure in the electron distribution function of small semiconducting structures: General features and specific trends*, Phys. Rev. **B-36** (1987), 1487; and *Ballistic electrons in a submicron structure: The distribution function and two-valley effects*, Physica B & C **134B** (1985), 470.

8. M. Rudan and F. Odeh, *Multi-dimensional discretization scheme for the hydrodynamic model of semiconductor equations*, COMPEL **5** (1986), 149.

9. R. K. Cook and J. Frey, *An efficient technique for two-dimensional simulation of velocity overshoot effects in Si and GaAs devices*, COMPEL **1** (1982); 65 and *Two-dimensional numerical simulation of energy transports effects in Si and GaAs*, MESFET's, IEEE Trans. on Electron Devices **ED-29** (1982), 970.

10. D. L. Rode, *Low-field electron transport*, in Semiconductors and Semimetals **10**, Academic Press, New York, 1975.

11. E. M. Conwell, *High field transport in semiconductors*, in Solid State Physics **9**, Academic Press, New York, 1967.

12. C. Kittel, *Introduction to solid state physics*, John Wiley, New York, 1968.

13. J. R. Chelikowski and M. L. Cohen, *Nonlocal pseudopotential calculations for the electronic structure of eleven diamond and zinc-blend semiconductors*, Phys. Rev. **B-14** (1976), 556.

14. J. M. Ziman, *Electrons and phonons*, Oxford University Press, London, 1960.

15. B. Niclot, P. Degond, and F. Poupaud, *Deterministic particle simulations of the Boltzmann equation of semiconductors*, J. Comput. Phys. **78** (1988), 313.

16. P. Lugli and D. K. Ferry, *Electron-electron interaction and high field transport in Si*, Appl. Phys. Lett. **46** (1985), 594; and M. Inoue and J. Frey, *Electron-electron interaction and screening effects in hot electron transport in GaAs*, J. Appl. Phys. **51** (1980), 4234.

17. J. J. Duderstadt and M. R. Martin, *Transport theory*, John Wiley, New York, 1979.

18. F. Poupaud, *Diffusion approximation and Milne problem for a Boltzmann equation of semiconductors*, Internal report n°150, CMAP, Ecole Polytechnique, Palaiseau (France), 1986.

19. ——, *On a system of nonlinear Boltzmann equations of semiconductor physics*, manuscript (to appear).

20. J. H. Fertziger and H. G. Kaper, *Mathematical theory of transport processes in gases*, North-Holland, Amsterdam, 1972.

21. A. H. Marshak and K. M. Van Vliet, *Electrical current in solids with position-dependent band structure*, Solid State Electron **21** (1978), 417.

22. E. W. Larsen and J. B. Keller, *Asymptotic solution of neutron transport problems for small mean free paths*, J. Math. Phys. **15** (1974), 75.

23. A. Bensoussan, J. L. Lions, and G. C. Papanicolaou, *Boundary layers and homogenization of transport processes*, J. Publ. RIMA Kyoto Univ. **15** (1979), 53–157.

24. C. Bardos, R. Santos, and R. Sentis, *Diffusion approximation and computation of the critical size*, Trans. Amer. Math. Soc. **284** (1984), 617.

25. W. Fawcett, A. D. Boardman, and S. Swain, *Monte-Carlo determination of electron transport properties in gallium arsenide*, J. Phys. Chem. Solids **31** (1970), 1963.

26. M. Mouis, J. F. Pone, and R. Castagne, *2-D particle modelling of the InP MISFET*, Solid State Electron **27** (1984), 659.

27. H. U. Baranger, *Ballistic electrons in a submicron semiconducting structure: A Boltzmann equation approach*, Ph.D. Thesis, Cornell University, 1986; See also Ref. 7.

28. S. Mas-Gallic, *A deterministic particle method for the linearized Boltzmann equation*, Transport Theory and Statist. Phys. **16** (1987), 855.

29. G. H. Cottet and S. Mas-Gallic, *A particle method to solve transport-diffusion equations, Part* 1: *the linear case*, Report n° 115, CMAP, Ecole Polytechnique, Palaiseau (France), 1983.

30. S. Mas-Gallic and P. A. Raviart, *Particle approximation of convection-diffusion problems*, report, University of Paris 6, Paris, France, 1985.

31. P. A. Raviart, *An analysis of particle methods*, in Numerical Methods in Fluid Dynamics, F. Brezzi, ed., Lecture Notes in Mathematics 1127, Springer-Verlag, Berlin, New York, 1985.

32. R. W. Hockney and J. W. Eastwood, *Computer simulations using particles*, McGraw-Hill, New York, 1981.

33. C. K. Birdsall and A. B. Langdon, *Plasma physics via computer simulations*, McGraw-Hill, New York, 1985.

34. G. H. Cottet and P. A. Raviart, *On particle-in-cell methods for the Vlasov-Poisson equations*, Transport Theory Statist. Phys. **15** (1986), 1.

35. ——, *Particle methods for the one-dimensional Vlasov-Poisson equations*, SIAM J. Numer. Anal. **21** (1984), 52.

36. S. Mas-Gallic and F. Poupaud, *Approximation of the transport equation by a weighted particle method*, Transport Theory Statist. Phys. **17** (1988), 311.

37. P. Degond and B. Niclot, *Numerical analysis of the weighted particle method applied to the semiconductor Boltzmann equation*, Numer. Math. **55** (1989), 599.

38. H. D. Rees, *Calculation of distribution functions by exploiting the stability of the steady state*, J. Phys. Chem. Solids **30** (1969), 643 and P. Kuivalainen and K. Lindberg, *Application of the integral Boltzmann equation to the hot electron problem in an inhomogeneous structure*, in High Speed Electronics by B. Küllback and H. Beneking (eds.), Springer-Verlag, New York, 1986.

Centre de Mathématiques Appliquées, Ecole Polytechnique 91128 Palaiseau Cedex

Lectures in Applied Mathematics
Volume **25** (1990)

The Transient Semiconductor Problem with Generation Terms

THOMAS I. SEIDMAN

Abstract. We consider a modified time-dependent van Roosbroeck system with saturating velocities and a source term subject to assumptions designed to permit a standard generation term corresponding to impact ionization. This is then a nonlinear parabolic system with a right-hand side of the form $S = S(\cdot, u, \nabla u, \mathscr{E})$, coupled with a map: $u \mapsto \mathscr{E}$ given by the Poisson equation at each t. The analysis leads to existence and uniqueness of solutions for all time. Comments are made as to such other considerations as temperature coupling, corner singularities, field-dependent diffusion, etc.

1. Introduction. We consider here a *drift/diffusion model* for semiconductor device physics. Our concern, not entirely anomalous even in the context of a seminar addressed specifically to *computational* aspects of semiconductor simulation, is with the fundamental mathematical question:

Is the problem well-posed?

In comparison with previous analyses of the time-dependent problem (e.g., [**2**], [**7**], [**5**]) we note the following significant differences.

• The *source* for the conservation law (continuity equations) admits *generation terms* as well as recombination—with hypotheses specifically formulated to permit the standard formulation $\alpha(|\mathscr{E}|)|\mathscr{J}|$ for impact ionization ([**8**], pp. 110ff.).

1980 *Mathematics Subject Classification* (1985 *Revision*). Primary 35K60, 35Q20, 78A35.

This research was partially supported by the Air Force Office of Scientific Research under grants #AFOSR-82-0271 and #AFOSR-87-0190.

• Drift velocities will be taken as saturating but otherwise fairly arbitrary ([**8**], pp. 93ff.).

• Temperature coupling is omitted—but this is only for expository simplification.

The structure of the resulting system is such that previously used methods (based on an essential use of maximum principle estimates for the concentrations) seem to be unavailable and new arguments are needed—including, for example, a new compactness result extending the Aubin Theorem [**1**] to the L^∞ case.

Certain obvious additional concerns, omitted here, will be addressed in a fuller presentation [**6**], now in preparation, and/or constitute an agenda for further investigation. It should be noted that the exposition here follows the presentation at the seminar, although some improved results have since been obtained. Some discussion of these considerations appears as the final section.

The plan of the paper is to discuss the formulation and hypotheses in the next section followed by a discussion of the "strategy" for the existence/uniqueness proof. The bulk of the paper is then a sketch of the proof itself, followed by a final section of further remarks, as noted above.

2. Formulation and hypotheses. The *drift/diffusion model* for semiconductor device physics consists of two parts: the conservation law for the concentrations and the Poisson equation for the electrostatic potential ψ, giving the field $\mathscr{E}$ as $-\nabla\psi$. We view these as defining a pair of coupled mappings, giving the set of concentrations in terms of an (assumed) field and giving the field in terms of the concentrations. Either of these—and thereby the other—can then be obtained as a fixed point of the composed maps.

A quite general conservation law takes the form:

$$\dot{u}_k - \nabla\cdot\mathscr{J}_k = S_k \quad \text{with } \mathscr{J}_k = D_k\nabla u_k + u_k\mathscr{V}_k \tag{2.1}$$

where $\mathbf{u} = (u_1, \ldots, u_K)$ is the set[1] of *concentrations*, and, suppressing the index k as we will do in general, $\mathscr{J}$ is the *current*, S is the *source*,

[1]For device physics one usually considers only two concentrations n and p which we might take to be u_1 and u_2, i.e., $K = 2$. The notation here, permitting a more general number K of concentrations, affords some mathematical generalization at no cost while simplifying the presentation and, indeed, this generalization may be of some interest. E.g., one might wish to consider a larger number of "species" in connection with process modelling and (perhaps a more interesting possibility) could take "electrons" as comprising several species corresponding to different bands.

D is the *diffusion coefficient*, and $\mathscr{V}$ is the *drift velocity*. The physics resides in the modelling of S, D, $\mathscr{V}$.

Of particular interest here is the modelling of S. We wish to be able to include standard models of recombination and also generation terms. For the latter, we take terms of the form: $\alpha(|\mathscr{E}|)|\mathscr{J}|$, modelling impact ionization, as typical of what we wish to permit. Comparing with the expression for $\mathscr{J}$ in (2.1), we see that this corresponds to considering[2]

$$S = S(\cdot, \mathbf{u}, \nabla\mathbf{u}, \mathscr{E})$$

of linear growth in $\mathbf{u}$, $\nabla\mathbf{u}$ and with suitable Lipschitz continuity. Thus we assume

$$|S(\cdots)| \le S^0 + \gamma[|\mathbf{u}| + |\nabla\mathbf{u}|], \tag{2.2}$$

with $S^0 \in L^2(\mathscr{Q})$ and

$$\begin{aligned} |S(\cdot, \mathbf{u}', \nabla\mathbf{u}', \mathscr{E}') &- S(\cdot, \mathbf{u}'', \nabla\mathbf{u}'', \mathscr{E}'')| \\ &\le \gamma[|\mathbf{u}' - \mathbf{u}''| + |\nabla(\mathbf{u}' - \mathbf{u}'')|] \\ &\quad + \gamma[1 + \max\{|\mathbf{u}'|, |\mathbf{u}''|, |\nabla\mathbf{u}'|, |\nabla\mathbf{u}''|\}]|\mathscr{E}' - \mathscr{E}''|. \end{aligned}$$

One easily sees that this is equivalent to the less symmetric form

$$\begin{aligned} |S(\cdot, \mathbf{u}', \nabla\mathbf{u}', \mathscr{E}') &- S(\cdot, \mathbf{u}'', \nabla\mathbf{u}'', \mathscr{E}'')| \\ &\le \gamma[|\mathbf{u}' - \mathbf{u}''| + |\nabla(\mathbf{u}' - \mathbf{u}'')|] \\ &\quad + \gamma[1 + \max\{|\mathbf{u}'|, |\nabla\mathbf{u}'|\}]|\mathscr{E}' - \mathscr{E}''| \end{aligned} \tag{2.3}$$

which we will use. Note that this gives (pointwise a.e. in t)

$$\|S' - S''\| \le \gamma[\|\mathbf{u}' - \mathbf{u}''\|_{(1)} + (1 + \|\mathbf{u}'\|_{(1)})\|\mathscr{E}' - \mathscr{E}''\|_\infty] \tag{2.4}$$

where $\|\cdot\|$ denotes the $L^2(\Omega)$-norm, $\|\cdot\|_{(1)}$ the $H^1(\Omega)$-norm and $\|\cdot\|_\infty$ the sup-norm on $\overline{\Omega}$.

The drift velocities $\mathscr{V} = \mathscr{V}_k$ are here not taken to grow linearly with the field $\mathscr{E}$ but, rather, we assume *velocity saturation*[3]; thus

$$|\mathscr{V}(\cdot, \mathscr{E})| \le \gamma. \tag{2.5}$$

---

[2]Note that we are here considering interaction terms so, of course, each $S = S_k$ must be taken to depend on $\mathbf{u}$ and $\nabla\mathbf{u}$ rather than just on the component u_k itself and ∇u_k.

[3]This seems to be a more physically realistic model than taking $\mathscr{V} = \pm\mu\mathscr{E}$ with a more-or-less constant mobility μ. On the other hand, the formulation and hypotheses here—using $\mathscr{V}$ directly, rather than introducing and modelling mobilities—leaves out much of the relevant physics: we will not even need to require any relations between the velocities and the field direction as were essential for the maximum principle arguments used heretofore. In some sense we eschew finesse for bounds. However, it is not clear how this aspect of the physics could be relevantly introduced in the present setting to improve the results.

We assume a uniform Lipschitz condition with respect to $\mathscr{E}$,

$$|\mathscr{V}(\cdot, \mathscr{E}') - \mathscr{V}(\cdot, \mathscr{E}'')| \le \gamma |\mathscr{E}' - \mathscr{E}''|. \tag{2.6}$$

Finally, we must consider the modelling of the diffusion coefficients $D = D_k$. It might well be more physically reasonable to take $D = D(\cdot, \mathscr{E})$—or even $D = D(\cdot, \mathscr{E}, \nabla \mathbf{u})$—but here we will consider only t, x dependence for D with

$$0 < \delta \le D(\cdot) \le \gamma \tag{2.7}$$

and with "sufficient regularity" (in t, x) that parabolic equations of the form:

$$\dot{v} - \nabla \cdot D \nabla v = g \quad + \text{[suitable initial and boundary conditions]}$$

can be expected to have weak solutions, estimable by the usual techniques.

Returning to consideration of the system of conservation equations (2.1), we adjoin boundary conditions of the form:

$$\begin{aligned} \mathscr{J}_k \cdot \mathbf{n} = \phi_k(\cdot) &\quad \text{on } \Gamma = \Gamma_N, \\ u_k = \hat{u}_k &\quad \text{on } \Gamma' = \Gamma_D, \end{aligned} \tag{2.8}$$

where $\phi_k = \phi_k(\cdot)$ is a given function on Γ and $\hat{u}_k$ is a given function on $\mathscr{Q}$ whose restriction to Γ' appears above. It is then convenient to impose the initial condition

$$u_k = \hat{u}_k \quad \text{on } \Omega \text{ at } t = 0, \tag{2.9}$$

using the same function $\hat{u}_k$. The weak formulation[4] of this part of the problem is then, suppressing the index k again,

$$\begin{aligned} \int_\Omega F(u - \hat{u})^{\cdot} + \int_\Omega \nabla F \cdot D \nabla (u - \hat{u}) \\ = \int_\Omega F(S - \widehat{S}) - \int_\Omega \nabla F \cdot u \mathscr{V} + \int_\Gamma F \hat{\phi} \end{aligned} \tag{2.10}$$

for any suitable *test function* F—meaning, essentially, for F vanishing on $\Gamma' = \partial\Omega \backslash \Gamma$—where we have taken $\widehat{S}$ and $\hat{\phi}$ so $\hat{u}$ satisfies

$$\hat{u}^{\cdot} - \nabla \cdot D \nabla \hat{u} = \widehat{S}, \qquad D \nabla \hat{u} \cdot \mathbf{n} = \phi - \hat{\phi} \quad \text{on } \Gamma.$$

We will assume that $\hat{u}$, $\widehat{S}$, and $\hat{\phi}$ are all "nice enough" for our purposes.

---

[4]Actually, we will always be using (2.10) integrated over time intervals $[0, t]$.

We conclude this section with a discussion of the treatment of the Poisson equation giving $\mathscr{E}$. From standard electrostatics one has[5]

$$(2.11)\qquad -\nabla\cdot\underline{\varepsilon}\nabla\psi = N + \mathbf{q}\cdot\mathbf{u} = N + \sum q_k u_k$$

where N is the intrinsic charge due to the doping. It is convenient, at this point, to change notation. Now writing $\mathscr{E}_{\text{total}}$ for the field previously denoted simply by $\mathscr{E}$, we introduce $\widehat{\mathscr{E}} = -\nabla\hat{\psi}$ where $\hat{\psi}$ is the solution of (2.11) with $\mathbf{u} = 0$: we then use the symbol $\mathscr{E}$ to denote the *difference* $\mathscr{E}_{\text{total}} - \widehat{\mathscr{E}}$. The point is that (with the new notation) $\mathscr{E}$ depends *linearly* on $\mathbf{u}$. We have $\mathscr{E} = -\nabla\psi$ where ψ now is the solution of the Poisson equation

$$(2.12)\qquad -\nabla\cdot\underline{\varepsilon}\nabla\psi = \mathbf{q}\cdot\mathbf{u}$$

with *homogeneous* boundary conditions[6] of the same form as those used for (2.11). Since $\widehat{\mathscr{E}}$ is fixed *a priori* and field differences are unaffected, we may assume that it is this *new* $\mathscr{E}$ which appears in the determination of S, $\mathscr{V}$. Clearly, there is now defined a linear map[7]

$$(2.13)\qquad \mathbf{L}\colon \mathbf{u}\mapsto\mathscr{E}$$

and our significant hypothesis[8] here is that (pointwise in t)

$$(2.14)\qquad \mathbf{L}\colon L^p(\Omega)\to\mathscr{C}(\overline{\Omega}) \text{ is compact}$$

for a suitable choice of p.

It is well-known ([**4**], p. 161) that—*under suitable regularity conditions* on $\underline{\varepsilon}$ *and* $\partial\Omega'$—we would have $\psi\in W^{2,p}$ for $\mathbf{u}\in L^p$. In that case one would have $\mathscr{E}\in W^{1,p}$ and this embeds compactly in $\mathscr{C}(\overline{\Omega})$ when p can be taken greater than the dimension, here assumed to be three. The unfortunate restriction here is on the regularity of the boundary: the

---

[5]More correctly, one should consider electrodynamics:

$$(1/c^2)\ddot{\psi} - \nabla\cdot\underline{\varepsilon}\nabla\psi = N + \mathbf{q}\cdot\mathbf{u}$$

but it is universally assumed that it is appropriate to take $c = \infty$ here. One thus has a family of elliptic equations on Ω, parametrized by t.

[6]Note that it need not be appropriate to consider this as holding for the same spatial region Ω as is considered for (2.1). It is, indeed, reasonable that this region Ω' contain Ω and we assume that $\mathbf{u}$, as it appears on the right here, is given *a priori* outside Ω. The boundary conditions for ψ are, of course, to be imposed at $\partial\Omega'$ and might include "conditions at infinity." The $\mathscr{E}$ to be used here is then the restriction of $-\nabla\psi$ to Ω.

[7]We will not distinguish notationally between $\mathbf{L}$ acting pointwise in t or for functions on $[0, T]$.

[8]We note that an equivalent formulation is that the embedding: $\mathscr{R}\hookrightarrow\mathscr{C}(\overline{\Omega})$ is compact where $\mathscr{R}$ is the range of the map $\mathbf{L}$ of (2.14), taken with graph norm.

kind of corners associated with reasonable device geometries are inconsistent with the assumption (2.14). For the present, however, we accept this restriction on the applicability of the results and impose (2.14).

Our formulation of the original Poisson equation (2.11) is now:

$$\mathscr{E}_{\text{total}} = \widehat{\mathscr{E}} + \mathbf{Lu} \tag{2.15}$$

with the data, i.e., the doping N and the voltages across the device, entirely absorbed in the specification of $\widehat{\mathscr{E}}$.

3. Strategy. The goal of this paper is the following theorem.

THEOREM 1. *In the setting and under the hypotheses of the previous section, the coupled system* (2.1), (2.11) *has a unique weak solution* $[\mathbf{u}^*, \mathscr{E}^*]$.

STRATEGY. The proof proceeds in several steps. Note, first, that our basic approach to the existence argument considers the system of conservation laws (2.10), with $\mathscr{E}$ temporarily fixed, as a nonlinear parabolic system determining $\mathbf{u}$. This defines a map:

$$\mathbf{T}_0 : \mathscr{E} \mapsto \mathbf{u} \tag{3.1}$$

once one shows that (2.10) has a unique solution for each $\mathscr{E}$. Setting $\mathbf{T}: \mathscr{E} \mapsto \mathbf{LT}_0(\mathscr{E})$, it is clear that any fixed point $\mathscr{E}^*$ of $\mathbf{T}$ provides a solution to the system, taking $\mathbf{u}^* = \mathbf{T}_0(\mathscr{E}^*)$.

1] (*Defining* $\mathbf{T}_0$). We seek $\mathbf{u}$ in the usual space[9]

$$\mathscr{X} = L^\infty(\to L^2) \cap L^2(\to H^1). \tag{3.2}$$

We will use exponentially weighted norms for $\mathscr{X}$,

$$\begin{aligned} \|f\|_{\mathscr{X}} &= \|f\|_{[r]} \\ &= [\sup\{e^{-rt}\|f(t)\|^2 : 0 < t < T\} + \delta \int_0^t e^{-rs}\|f(s)\|^2_{(1)}]^{1/2}, \end{aligned} \tag{3.3}$$

where r will be specified later. The argument is by consideration of another fixed-point problem, using the standard Picard iteration and the Contractive Mapping Theorem[10]. This is, indeed, standard enough for

[9]This notation is a short form of

$$\mathscr{X} = L^\infty([0, T] \to L^2(\Omega)) \cap L^2([0, T] \to H^1(\Omega))$$

with the domains suppressed. We will not explicitly distinguish, here and elsewhere, between spaces of scalar- or vector-valued functions.

[10]It is the use of an exponentially weighted norm for $\mathscr{X}$ which makes it possible to get estimates directly for an arbitrary time interval $[0, T]$.

us to have simply cited a reference but provides the simplest occasion to show the kind of estimates we will be using. The relevant computation is Estimate 1 for differences, using the Lipschitz condition (2.4) with $\mathscr{E}$ fixed.

2] (*A priori bounds*). We obtain an estimate in $\mathscr{X}$ for $\mathbf{u} = \mathbf{T}_0(\mathscr{E})$. This will be Estimate 2, using the "growth condition" (2.2) as well as (2.5), (2.7). Since (2.2) does not involve $\mathscr{E}$, the estimate obtained is independent of the particular $\mathscr{E}$ used. For later use we note that this also bounds S in $L^2(\mathscr{Q})$ and, using the equation (2.1), bounds $\dot{u}$ in $L^2(\to H^{-1})$. Observing that use of (2.13), (2.14) pointwise in t will require a pointwise L^p-bound on $\mathbf{u}$, we bound $w = |u|^\lambda$ $(\lambda = p/2)$ in $\mathscr{X}$. Again, this Estimate 3 uses (2.2), etc., and gives a bound, independent of $\mathscr{E}$, for $\mathbf{u}$ in $L^\infty(\to L^p)$.

3] (*Continuity and compactness*). A computation, Estimate 4 for differences, using the Lipschitz conditions (2.3), (2.6), gives continuity of $\mathbf{T}_0\colon L^\infty(\mathscr{Q}) \to \mathscr{X}$.

Estimates 2 and 3, with the use of the assumption (2.14) and the uniform bound noted above for $\dot{\mathbf{u}}$, suffice to show that the range of $\mathbf{T}$, i.e., the set

$$\mathscr{K}_{\mathscr{E}} := \{\mathbf{Lu}\colon \mathbf{u} = \mathbf{T}_0(\mathscr{E}) \text{ for some } \mathscr{E}\}, \tag{3.4}$$

is precompact in $\mathscr{C}(\overline{\mathscr{Q}}) = \mathscr{C}([0, T] \to \mathscr{C}(\overline{\Omega}))$. The argument involves an extension of the Aubin Compactness Theorem to the L^∞ case; we state that here without proof.

THEOREM 2. *Let $\mathscr{X}_1 \hookrightarrow \mathscr{X} \hookrightarrow \mathscr{X}_2$ be Banach spaces such that the embedding $\mathscr{X}_1 \hookrightarrow \mathscr{X}$ is compact. For a compact interval $\mathscr{J}$ in* $\mathbf{R}$, *consider a set $\mathscr{K} \subset L^\infty(\mathscr{J} \to \mathscr{X}_1)$ such that $\{\dot{x}(\cdot)\colon x \in \mathscr{K}\}$ is bounded in $L^p(\mathscr{J} \to \mathscr{X}_2)$ for some $p > 1$. Then $\mathscr{K}$ is precompact in $C(\mathscr{J} \to \mathscr{X})$.*

One easily sees that $\mathbf{L}$ has closed graph in $\mathscr{X} \times C(\overline{\mathscr{Q}})$ which, with the compactness of $\mathscr{K}_{\mathscr{E}}$ and continuity of $\mathbf{T}_0$, gives continuity of the composition $\mathbf{T} = \mathbf{LT}_0\colon C(\overline{\mathscr{Q}}) \to C(\overline{\mathscr{Q}})$.

4] (*Existence*). Consider the compact convex set

$$\mathscr{K}_{\mathscr{E}}^* := \overline{\text{co}}[\mathscr{K}_{\mathscr{E}}]. \tag{3.5}$$

Clearly this is invariant under $\mathbf{T}$ and, with the continuity noted above, the Schauder Fixed Point Theorem applies to gives existence of a fixed point of $\mathbf{T}$ and so a solution of the coupled system.

5] (*Uniqueness*). Assume two, putatively distinct, solutions $\mathbf{u}'$, $\mathbf{u}''$ and estimate the difference $\mathbf{u} = \mathbf{u}' - \mathbf{u}''$. The argument involves analysis of the coupling of Estimate 5 for $\mathbf{u} \in \mathscr{X}$ and Estimate 6 for $\mathbf{u} \in L^\infty(\to L^p)$ to show $\mathbf{u} = \mathbf{0}$.

In the next two sections we indicate roughly how the estimates which constitute this proof are to be obtained.

4. Existence. In this section we go through the first three items of the strategy described in the previous section, thus sketching a proof of existence of a fixed point of $\mathbf{T}$ and of a solution of the system (2.1), (2.11).

1] *Fix* $\mathscr{E}$. For any $\bar{\mathbf{u}} \in \mathscr{X}$ there is a (unique) weak solution $u = u_k$ of each of the *linear* parabolic equations (2.1), (2.8), using $\mathscr{V} = \mathscr{V}_k(\cdot, \mathscr{E})$ with the given $\mathscr{E}$ and $S = S_k = S(\cdot, \bar{\mathbf{u}}, \nabla\bar{\mathbf{u}}, \mathscr{E})$. We show that the induced Picard map: $\bar{\mathbf{u}} \mapsto \mathbf{u}$ is contractive on $\mathscr{X}$ by a standard argument. Hence there is a unique fixed point $\mathbf{u} = \mathbf{T}_0(\mathscr{E})$.

Given $\bar{\mathbf{u}}'$ and $\bar{\mathbf{u}}''$, let $u' = u'_k$, and $u'' = u''_k$ be the corresponding solutions of (2.1), (2.8) using S' and S''. Set $u = u' - u''$, $\bar{\mathbf{u}} = \bar{\mathbf{u}}' - \bar{\mathbf{u}}''$, etc.; note that $\mathbf{u}(0) = \mathbf{0}$. We consider the equations in weak form as (2.10) and take the difference, obtaining

$$\int_\Omega F\dot{u} + \int_\Omega \nabla F \cdot D\nabla u = \int_\Omega F(\bar{S}' - \bar{S}'') - \int_\Omega \nabla F \cdot u\mathscr{V}, \tag{4.1}$$

with no boundary term since the data match. We will take

$$F = 2e^{-rt}u \quad \text{so } F\dot{u} = [e^{-rt}u^2]^{\cdot} + re^{-rt}u^2 \tag{4.2}$$

with r to be determined. Integrating (4.1) on $[0, t]$ then gives[11]

$$\begin{aligned} e^{-rt}\|u(t)\|^2 + (r - \delta)\int_0^t e^{-rs}\|u\|^2 + 2\delta\int_0^t e^{-rs}\|u\|^2_{(1)} \\ \leq M\int_0^t e^{-rs}\|u\|\,\|\bar{\mathbf{u}}\|_{(1)} + \gamma\int_0^t e^{-rs}\|u\|_{(1)}\|u\| \\ \leq \delta\int_0^t e^{-rs}\|u\|^2_{(1)} + M\int_0^t e^{-rs}\|u\|^2 + \frac{\delta}{2K}\int_0^t e^{-rs}\|\bar{\mathbf{u}}\|^2_{(1)}. \end{aligned} \tag{4.3}$$

Requiring now that $r > M + \delta$ and summing over the K components, one obtains

$$\|\mathbf{u}\|^2_{\mathscr{X}} \leq \tfrac{1}{2}\|\bar{\mathbf{u}}\|^2_{\mathscr{X}} \tag{Estimate 1}$$

[11]Just as "γ" was used generically in the growth and Lipschitz conditions of the hypotheses, we use "M" here and later for a generic constant—different in different occurrences—which may depend only on γ, δ, Ω, K, and the parameters of (2.14).

which, of course, is just the desired contractivity.

2] *Set* $z = u_k - \hat{u}_k$, *etc.* For Estimate 2 we take

$$F = 2e^{-rt} z \quad \text{so } F\dot{z} = [e^{-rt} z^2]^{\cdot} + re^{-rt} z^2 \tag{4.4}$$

in (2.10) and again integrate on $[0, t]$. We now get

$$\begin{aligned} e^{-rt}\|z(t)\|^2 + (r-\delta)\int_0^t e^{-rs}\|z\|^2 + 2\delta\int_0^t e^{-rs}\|z\|_{(1)}^2 \\ \le \delta\int_0^t e^{-rs}\|z\|_{(1)}^2 + M\int_0^t e^{-rs}\|z\|^2 \\ + \frac{\delta}{2K}\int_0^t e^{-rs}\|z\|_{(1)}^2 \\ + M'\int_0^t e^{-rs}[\|S^0\|^2 + \|\hat{\mathbf{u}}\|_{(1)}^2 + \|\hat{\phi}\|_\Gamma^2] \end{aligned} \tag{4.5}$$

where we have treated the boundary term by using compactness of the trace: $H^1(\Omega) \hookrightarrow L^2(\Gamma)$ whence

$$\|z\|_\Gamma^2 \le \varepsilon\|z\|_{(1)}^2 + C_\varepsilon\|z\|^2. \tag{4.6}$$

Summing over k and taking $r > M + \delta$ (4.5) provides a uniform bound in $\mathscr{X}$ for $\mathbf{u} = \mathbf{T}_0(\mathscr{E})$, i.e., a bound[12] independent of $\mathscr{E}$.

To bound $\mathbf{u}$ in $L^\infty(\to L^p)$, we introduce

$$v := |z|^{p-1}\operatorname{sgn} z, \qquad w := |z|^\lambda \quad (p = 2\lambda) \tag{4.7}$$

and set $F = pe^{-rt}v$ in (2.10). We note the identities:

$$\begin{aligned} zv &= w^2 = |z|^p, \\ |z\nabla v| &= 2(1 - 1/p)|w|\,|\nabla w|, \\ |v\nabla z| &= (2/p)|w|\,|\nabla w|, \\ |\nabla v \cdot \nabla z| &= [4(p-1)/p^2]|\nabla w|^2 \\ &\qquad (\text{set } \delta' = 4\delta(p-1)/p), \\ F\dot{z} &= [e^{-rt}w^2]^{\cdot} + re^{-rt}w^2. \end{aligned} \tag{4.8}$$

We also note that, assuming $p < 6$, we have

$$\|v\| \le M\|w\|^{1-1/p}\|w\|_{(1)}, \qquad \|v\|_\Gamma \le M\|w\|_{(1)}^{2-2/p}. \tag{4.9}$$

Since the computation for this Estimate 3 is similar to that for the Estimate 6 in the next section, we note here only that this leads to an $L^\infty(\to L^p)$ bound, independent of $\mathscr{E}$, for $\mathbf{u}$.

---

[12]This estimate was obtained originally during a visit with Peter Markowich at Tech. Univ. Wien. The feeling then—now justified!—was, "Given such an estimate, there *must* be a theorem somewhere!"

3]. For Estimate 4 we consider $\mathbf{u}' = \mathbf{T}_0(\mathscr{E}')$, $\mathbf{u}'' = \mathbf{T}_0(\mathscr{E}'')$ and set $u = u' - u''$, etc., much as for Estimate 1; instead of (4.1), one now has

$$\int_\Omega F\dot{u} + \int_\Omega \nabla F \cdot D\nabla u = \int_\Omega F(S' - S'') - \int_\Omega \nabla F \cdot u\mathscr{V}' - \int_\Omega \nabla F \cdot u''(\mathscr{V}' - \mathscr{V}'') \tag{4.10}$$

and proceeds much as for (4.3). Using (2.4), (2.6), etc., one obtains

$$\begin{aligned} e^{-rt}\|u(t)\|^2 + (r-\delta)\int_0^t e^{-rs}\|u\|^2 + 2\delta\int_0^t e^{-rs}\|u\|^2_{(1)} \\ \leq \delta\int_0^t e^{-rs}\|u\|^2_{(1)} + M\int_0^t e^{-rs}\|u\|^2 \\ + M'\|\mathscr{E}' - \mathscr{E}''\|^2_\infty \int_0^t e^{-rs}\sigma^2 \end{aligned} \tag{4.11}$$

where $\sigma(t) = 1 + \|\mathbf{u}'\|_{(1)}$ and we have also used $\|\cdot\|_\infty$ for the sup-norm on $\overline{\mathscr{Q}}$ to be able to take this outside the integral. Thus $\mathbf{T}_0: \mathscr{C}(\overline{\mathscr{Q}}) \to \mathscr{X}$ is Lipschitzian on taking r suitably large in (3.3), (4.2). The remainder of the argument for the precompactness of $\mathscr{K}_\mathscr{E}$ and the continuity of $\mathbf{T}$ is straightforward, using Theorem 2.

4]. As noted in the previous section, the existence of a fixed point of $\mathbf{T}$ is now immediate from the Schauder Theorem, concluding the proof of the existence part of Theorem 1.

5. Uniqueness. To show the uniqueness of the solution whose existence in

$$L^2([0,T] \to H^1(\Omega)) \cap L^\infty([0,T] \to L^p(\Omega))$$

has just been demonstrated, we assume the existence of a pair of such solutions, $\mathbf{u}'$ and $\mathbf{u}''$ with corresponding $\mathscr{E}'$ and $\mathscr{E}''$ in $\mathscr{C}(\overline{\mathscr{Q}})$. We set

$$\begin{aligned} \mathbf{u} = \mathbf{u}' - \mathbf{u}'', \qquad \mathscr{E} = \mathscr{E}' - \mathscr{E}'' = \mathbf{L}\mathbf{u}, \\ \sigma = (1 + \|\mathbf{u}'\|_{(1)}) \in L^2(0,T). \end{aligned} \tag{5.1}$$

Now, essentially as for Estimate 4, we obtain

$$\begin{aligned} e^{-rt}\|u(t)\|^2 + (r-\delta)\int_0^t e^{-rs}\|u\|^2 + 2\delta\int_0^t e^{-rs}\|u\|^2_{(1)} \\ \leq \delta\int_0^t e^{-rs}\|u\|^2_{(1)} + M\int_0^t e^{-rs}\|u\|^2 \\ + M'\int_0^t e^{-rs}\sigma^2\|\mathscr{E}\|^2_\infty \end{aligned} \tag{5.2}$$

whence, taking r large enough and summing over components,

$$\text{(Estimate 5)} \qquad \int_0^t e^{-rs}\|\mathbf{u}\|_{(1)}^2 \le M \int_0^t e^{-rs}\sigma^2\|\mathscr{E}\|_\infty^2.$$

It is now convenient to set

$$\begin{aligned} \zeta(t) = \zeta_k(t) &= \|u_k(t)\|_p \\ &= \|w(t)\|^{1/\lambda} \quad (w = |u_k|^\lambda), \\ \tilde{\zeta}(t) &= \sup\{\zeta(s) \colon 0 \le s \le t\}, \\ \hat{\zeta}(t) &= \left[\sum_k \zeta_k^2(t)\right]^{1/2}. \end{aligned} \tag{5.3}$$

Note that $\|w\|^2 = \zeta^p$ and that (2.14) gives $\|\mathscr{E}\|_\infty \le M\hat{\zeta}$.

We now proceed as for Estimate 3, using (4.7), (4.8) with $z = u$. Using $F = pe^{-rt}v$ in (4.10) and integrating over $[0, T]$, one obtains, with the aid of (4.9),

$$\begin{aligned} e^{-rt}\|w(t)\|^2 &+ (r - \delta')\int_0^t e^{-rs}\|w\|^2 + \delta'\int_0^t e^{-rs}\|w\|_{(1)}^2 \\ &\le M\int_0^t e^{-rs}[\|w\|\,\|w\|_{(1)} + \|v\| + \|\mathbf{u}\|_{(1)}\|v\| + \sigma\|\mathscr{E}\|_\infty\|v\|] \\ &\le \delta'\int_0^t e^{-rs}\|w\|_{(1)}^2 + M\int_0^t e^{-rs}\zeta^p \\ &\quad + M'\int_0^t e^{-rs}[\|\mathbf{u}\|_{(1)}^2 + \sigma^2\|\mathscr{E}\|_\infty^2]\tilde{\zeta}^{p-2}. \end{aligned} \tag{5.4}$$

One easily sees that this gives

$$e^{-rt}\tilde{\zeta}^2(t) \le M\int_0^t e^{-rs}[\|\mathbf{u}\|_{(1)}^2 + \sigma^2\|\mathscr{E}\|_\infty^2]$$

whence, summing over components,

$$\text{(Estimate 6)} \qquad e^{-rt}\hat{\zeta}^2(t) \le \int_0^t e^{-rs}[\|\mathbf{u}\|_{(1)}^2 + \sigma^2\hat{\zeta}^2].$$

We substitute Estimate 5 into Estimate 6 and apply the Gronwall Inequality, noting the integrability of σ^2, to see that $e^{-rt}\hat{\zeta}^2$ vanishes; so $\mathbf{u}'$ and $\mathbf{u}''$ must be the same. We have thus shown that the solution is unique as asserted, completing the proof of Theorem 1.

6. Further remarks. We conclude with a number of comments on the continuation and extension of the work presented here.

REMARK 1. We could have included, with only trivial modification of the arguments presented here, a possible dependence of each of the drift velocities $\mathscr{V}_k$ on the corresponding density gradient[13]:

$$\mathscr{V}_k = \mathscr{V}_k(\cdot, \mathscr{E}, \nabla u_k),$$

although this could not be done so easily if one wished dependence on $\nabla \mathbf{u}$ instead.

We note also that, as in [7], by including the obvious structural hypotheses on $S(\cdots)$ and the data, one can ensure the requisite positivity of the concentrations u_k.

REMARK 2. Probably the most restrictive assumption made here is (2.14), which precludes consideration of geometries involving the possibility of gradient "corner singularities" for the Poisson equation. It is well known [3] that such singularities do occur—e.g., at re-entrant angles in two dimensions. Although the relevant theory is unavailable (especially for three-dimensional geometries), one would reasonably expect a general result of the form:

$$\mathbf{L}\colon \mathbf{u} \mapsto \mathscr{E}\colon L^p(\Omega) \to \mathscr{C}_\rho(\overline{\Omega}) \text{ is compact} \tag{6.1}$$

(for $p > 3$) as a replacement for (2.14). Here ρ is a weight function vanishing suitably near the "danger points" and $\mathscr{C}_\rho$ uses a ρ-weighted sup-norm. The arguments presented here would work if one could then estimate $\nabla \mathbf{u}$ with a $(1/\rho)$-weighted L^2-norm. In particular, if one had good information regarding the growth of $1/\rho$, then imposing stronger regularity conditions on D might give some $L^{\bar{p}}(\Omega)$ estimate for $\nabla \mathbf{u}$ which could suffice.[14]

REMARK 3. A result obtained since the talk is that one can consider

$$D = D(\cdot, \mathscr{E}, \mathbf{u}, \nabla \mathbf{u}), \qquad V = V(\cdot, \mathscr{E}, \mathbf{u}, \nabla \mathbf{u}), \tag{6.2}$$

as well as the replacement of (2.14) by (6.1), and show existence by application of the Schauder Theorem. No uniqueness argument is (yet) available for this more general context. It is also possible to include, here, such possibilities as having the Dirichlet data dependent on $\mathscr{E}$.

One can also include temperature coupling in the system. Thus, one would admit suitable dependence on the temperature ϑ for D, $\mathscr{V}$, S

[13] Compare the formulation in ([8], §4.1).

[14] It seems of interest, in any case, to consider the possibility of obtaining such improved regularity for the solution under suitably strengthened hypotheses on D—e.g., piecewise spatial continuity.

and, much as in [7], adjoin another parabolic equation for ϑ:

$$\dot{\vartheta} - \nabla \cdot C\nabla\vartheta = \mathscr{E}_{\text{total}} \cdot \left(\sum_k q_k \mathscr{J}_k\right). \tag{6.3}$$

In this context one still can show existence of a solution, now consisting of $[\mathbf{u}^*, \mathscr{E}^*, \vartheta^*]$ as functions on $\mathscr{Q}$.

REMARK 4. At the seminar, Hans Weinberger raised the question of whether, in view of the availability of Estimates 5 and 6, it might not be possible to show that $\mathbf{T}$ is actually contractive, thus providing a *constructive* argument for existence. The argument, as presented above, does seem to require the use of a single function σ. From the results obtained here it follows that $\mathscr{U} = \mathbf{T}_0(\mathscr{K}_{\mathscr{E}}^*)$ is compact and, using this, it has been possible to modify the use of Estimates 5 and 6 to show the contractivity of a related map with respect to a suitably constructed metric.

REMARK 5. As a further consideration for future work, it can be fairly confidently expected that "uniform" versions of these estimates (in particular, along the lines of the preceding remark) can be obtained for a suitable family of discretizations. This would provide a convergence proof for computational approximation, thus completing the justification of relevance of this material to the titular topic of the seminar.

REFERENCES

1. J. P. Aubin, *Un théorème de compacité*, C. R. Acad. Sci. Paris Sér. I Math. **265** (1963), 5042–5045.
2. H. Gajewski and K. Gröger, *On the basic equations for carrier transport in semiconductors*, J. Math. Anal. Appl. **113** (1986), 12–35.
3. P. Grisvard, *Elliptic problems in nonsmooth domains*, Pitman APP, Boston, 1985.
4. C. B. Morrey, Jr., *Multiple integrals in the calculus of variations*, Springer-Verlag, New York, 1966.
5. T. I. Seidman, *Time-dependent solutions of a nonlinear system arising in semiconductor theory*, II: *Boundedness and periodicity*, Nonlinear Anal. **10** (1986), 491–502.
6. ——, *Time-dependent solutions of a diffusion-reaction system with electrostatic convection*.
7. T. I. Seidman and G. M. Troianniello, *Time-dependent solutions of a nonlinear system arising in semiconductor theory*, Nonlinear Anal. **9** (1985), 1137–1157.
8. S. Selberherr, *Analysis and simulation of semiconductor devices*, Springer-Verlag, Wien, 1984.

DEPARTMENT OF MATHEMATICS AND STATISTICS
UNIVERSITY OF MARYLAND BALTIMORE COUNTY
CATONSVILLE, MARYLAND, 21228
(BITNET: SEIDMAN@UMBIC, INTERNET: SEIDMAN@UMBC1.UMBC.EDU)

Lectures in Applied Mathematics
Volume **25** (1990)

Modelling Gallium Arsenide Transistors

JAMES L. BLUE AND CHARLES L. WILSON

Abstract. The usual equations used in modelling semiconductor devices are a set of three coupled partial differential equations first formulated forty years ago. These PDEs are adequate to model present-day silicon transistors, but are inadequate for modelling GaAs transistors. Suitable PDEs can be formulated using quantum mechanics and statistical thermodynamics models obtained by Monte-Carlo calculations.

1. Introduction. For the last twenty years, many investigators have been modelling semiconductor devices using a set of three coupled partial differential equations first formulated forty years ago. These PDEs are a simplification of the actual quantum-mechanical/statistical-thermodynamical system. Comparison of experiment and calculation indicates that these PDEs are adequate to model silicon transistors with dimensions as small as 0.1 micrometers.

Gallium arsenide has a more complex band structure than silicon. The usual PDEs are inadequate for modelling GaAs transistors, as they are for modelling Si transistors at sufficiently high fields and for sufficiently small structures. Suitable PDEs can be formulated using quantum mechanics and statistical thermodynamics models obtained by Monte-Carlo calculations.

2. The usual derivation of the standard equations. The "standard" equations are obtained by treating the electron density as a simple charged fluid [**1**]. (Refer to [**2**] for a review of semiconductor device

1980 *Mathematics Subject Classification* (1985 *Revision*). Primary 82A30, 81H20, 35J60.

physics.) Conservation of electron density gives

$$\frac{\partial n}{\partial t} = G + \frac{1}{q}\nabla \cdot \mathbf{J}_n,$$

$$\mathbf{J}_n = -qn\mathbf{v}_n + qD_n\nabla n.$$

The assumption is made that the electron velocity may be found from the electric field using a mobility, $\mathbf{v}_n = -\mu_n \mathbf{E} = \mu_n \nabla\phi$. The Einstein relation, which is true in thermodynamic equilibrium, is assumed true, also: $D_n = \mu_n kT/q$. Thus the current density is

$$\mathbf{J}_n = q\mu_n(-\nabla\phi + (kT/q)\nabla n).$$

There is an obvious change of variable that gives a simplification. Define ϕ_n by $n = n_i \exp[(\phi - \phi_n)q/(kT)]$. Then the current density simplifies to $\mathbf{J}_n = q\mu_n\nabla\phi_n$, and the current continuity equation becomes

$$\frac{\partial n}{\partial t} = G - \nabla \cdot (\mu_n n \nabla\phi_n).$$

The derivation for the holes is similar.

In addition, we have Poisson's equation:

$$\nabla \cdot (\kappa\nabla\phi) = -\frac{q}{\varepsilon_0}(N - n).$$

These equations have been used for Si devices for years. They are often referred to as the drift and diffusion model, or the convection-diffusion model. Using these equations, it is possible to (1) prove some useful properties of the equations [**3**]; (2) solve the equations numerically [**4**] [**5**] [**6**]; (3) get solutions that agree with experiment [**7**].

3. A proper derivation of the standard equations. The "standard" equations are wrong; their derivation is over-simplified. The electrons are not a simple charged fluid. Fortunately, the equations are adequate for silicon devices, at least if the minimum horizontal dimensions of the devices are at least 0.1 micrometers, and the electric fields are not too large.

To derive the equations properly, we must go back to fundamentals. (This derivation follows that in [**8**], at least in spirit.) This means quantum mechanics, Schrödinger's equation for the wavefunction Ψ for many, many, many electrons in a perfect crystal.

We assume electrons do not interact with each other, but only with the crystal; this means that the electron density cannot be too large. Then we have "one-electron" wave functions: $\Psi = \prod \psi_j(\mathbf{r}_j, \mathbf{k}_j)$, where, according to the Bloch–Floquet theorem, $\psi(\mathbf{r}, \mathbf{k}) = \exp(i\mathbf{k} \cdot \mathbf{r})U(\mathbf{r}, \mathbf{k})$, where $\mathbf{k}$ ranges over the Brillouin zone. Each state has an energy $\mathscr{E}(\mathbf{k})$; the allowed energies lie in bands. Near a band minimum, $\mathscr{E}(\mathbf{k}) \approx \hbar^2 k^2/(2m^*)$, except the bands are really ellipsoidal, not spherical. Electrons fill the allowed states up to a certain energy, called the Fermi energy. In silicon at normal applied fields, only one region, or "valley," of the Brillouin zone contains significant numbers of electrons. In gallium arsenide, electrons can reach both the central valley and an outer, or satellite, valley.

$U(\mathbf{r}, \mathbf{k})$ has the same periodicity as the crystal lattice. The velocity of an electron is $\mathbf{v}(\mathbf{k}) = \nabla_{\mathbf{k}}\mathscr{E}(\mathbf{k})/\hbar$. An applied force changes $\mathbf{k}$; $\mathbf{k}$ obeys $\hbar d\mathbf{k}/dt = \mathbf{F}$.

We now leave quantum mechanics, and go to semiclassical physics, to statistical thermodynamics. We treat the motions of large numbers of electrons statistically. It is impossible to treat all the electrons in a device individually. Instead, consider a statistical distribution of electrons. There is a distribution function $f(\mathbf{r}, \mathbf{k}, t)$ for one electron, which obeys the Boltzmann equation,

$$\frac{\partial f}{\partial t} = -\frac{\partial \mathbf{k}}{\partial t} \cdot \nabla_{\mathbf{k}} f - \frac{\partial \mathbf{r}}{\partial t} \cdot \nabla_{\mathbf{r}} f + (\text{scattering terms}).$$

An observable quantity is an expectation value of f, and is a function of $\mathbf{r}$. It comes from an integral over $\mathbf{k}$ over the Brillouin zone. For example, the electron density is $n = \int f\, d\mathbf{k}$.

Integrating the Boltzmann equation over $\mathbf{k}$, we get

$$\frac{\partial n}{\partial t} - \frac{1}{q}\nabla \cdot \mathbf{J}_n = (\text{scattering terms}).$$

However, $\mathbf{J}_n$ is now defined by an expectation value,

$$\mathbf{J}_n = -\frac{1}{q}\int \mathbf{v} f\, d\mathbf{k}$$

and the velocity is $\mathbf{v}(\mathbf{k}) = \nabla_{\mathbf{k}}\mathscr{E}(\mathbf{k})/\hbar$.

We note that, at low values of $|\mathbf{E}|$, and under some conditions [**9**], the current density can be written as

$$\mathbf{J}_n = -qn\mathbf{v}_n + qD_n \nabla n.$$

Moments of the Boltzmann equation can be defined:

$$n = M_0 = \int f\, d\mathbf{k} \equiv \langle 1; f \rangle,$$

$$M_1 = \int \mathbf{v} f\, d\mathbf{k} \equiv \langle \mathbf{v}; f \rangle,$$

$$M_2 = \int \mathbf{v}\mathbf{v} f\, d\mathbf{k} \equiv \langle \mathbf{v}\mathbf{v}; f \rangle.$$

The standard equations use M_0 only. This is, in some sense, equivalent to assuming a simple form for the distribution function, f.

There are several possibilities to try to solve the Boltzmann equation, including: (1) Linearize, and maybe solve the linearized equation analytically. This is not useful for semiconductor devices. (2) Use the first few moments of the Boltzmann equation to generate PDEs. (3) Use Monte-Carlo methods to find f as a function of **r** and **k**. (4) A combination: use Monte-Carlo to estimate f as a function of a few variables, and use the moment equations for these variables. This method is used in the current work.

Let us now consider how to approximate f. With no fields,

$$f_0(\mathscr{E}) = \frac{1}{1 + e^{(\mathscr{E} - \mathscr{E}_f)/(kT)}}.$$

A general expression for the solution to the Boltzmann equation without scattering is

$$f_1(\mathscr{E}) = \frac{1}{1 + e^{(\mathscr{E} - q\phi + q\phi_n - \mathscr{E}_f)/(kT)}}.$$

Here $\mathscr{E}(\mathbf{k}) - q\phi$ is the energy of the electron, and $q\phi_n$ is a "chemical potential".

If $f = f_1 +$ (small terms), it may be shown that $\mathbf{J}_n = n\mu_n \nabla \phi_n$, and that $n = \int f_1(\mathscr{E}(\mathbf{k}))\, d\mathbf{k}$. Changing variables of integration to $\mathscr{E}$, we find that $n \sim \int f_1(\mathscr{E})\mathscr{E}^{1/2}\, d\mathscr{E}$;

$$n = N_c \mathscr{F}_{\frac{1}{2}}((q\phi - q\phi_n - \mathscr{E}_c), T)$$

where $F_{\frac{1}{2}}$ is the Fermi–Dirac integral of order one-half,

$$\mathscr{F}_{\frac{1}{2}}(\mathscr{E}, T) = \int_0^\infty \frac{1}{1 + e^{(z - \mathscr{E})/(kT)}}\, dz.$$

If the first argument of $\mathscr{F}_{\frac{1}{2}}$ is negative enough, then a good approximation is $n \approx n_i \exp[q(\phi - \phi_n)/(kT)]$. However, in all practical devices, there is always a region where the latter approximation is invalid, and where the electric field is too large for some of the other approximations.

Note that the quasi-Fermi level, ϕ_n, is not just a convenient change of variable.

4. Silicon vs. gallium arsenide. In GaAs, three different "kinds" of electrons are important, one for the central valley, one for the lower group of equivalent satellite valleys, and one for the upper group of equivalent satellite valleys. Electrons scatter from one kind to the other, as well as within a valley. (In silicon, inter-valley scattering is not important until very high electric fields.) We need to treat each kind separately, to give each valley its own distribution function. The total f is complicated, but the f in each valley is much simpler. Treating only the central valley and the lower group of satellite valleys limits the allowable maximum electric field somewhat, but saves on computer time.

How else are silicon and gallium arsenide different? Following is a graph of electron velocity vs. electric field in GaAs, obtained from a Monte-Carlo calculation [**8**].

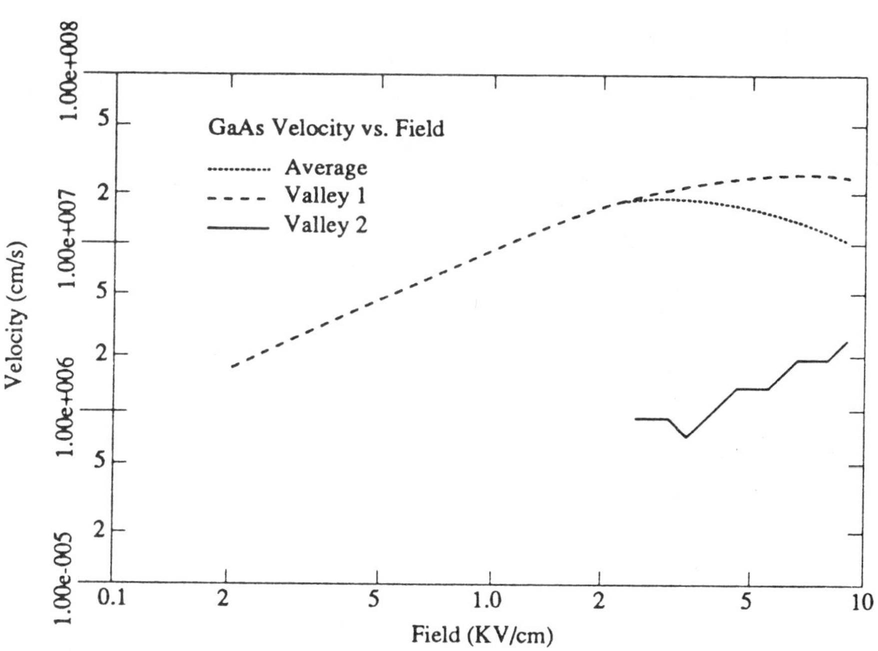

FIGURE 1

The drop in velocity at high fields in GaAs occurs because the central valley electrons scatter into the satellite valleys, where they move much more slowly.

To approximate f in each valley, we consider the Boltzmann equation with scattering terms. We assume uniform GaAs and a uniform electric field. We let one electron loose, follow its scatterings in the crystal, including from central valley to satellite valley, and keep track of time spent in various $\mathbf{k}$ states. The amount of time is proportional to $f(\mathbf{k})$. This gives $f(\mathbf{k})$ independent of $\mathbf{r}$, but dependent on $|\mathbf{E}|$.

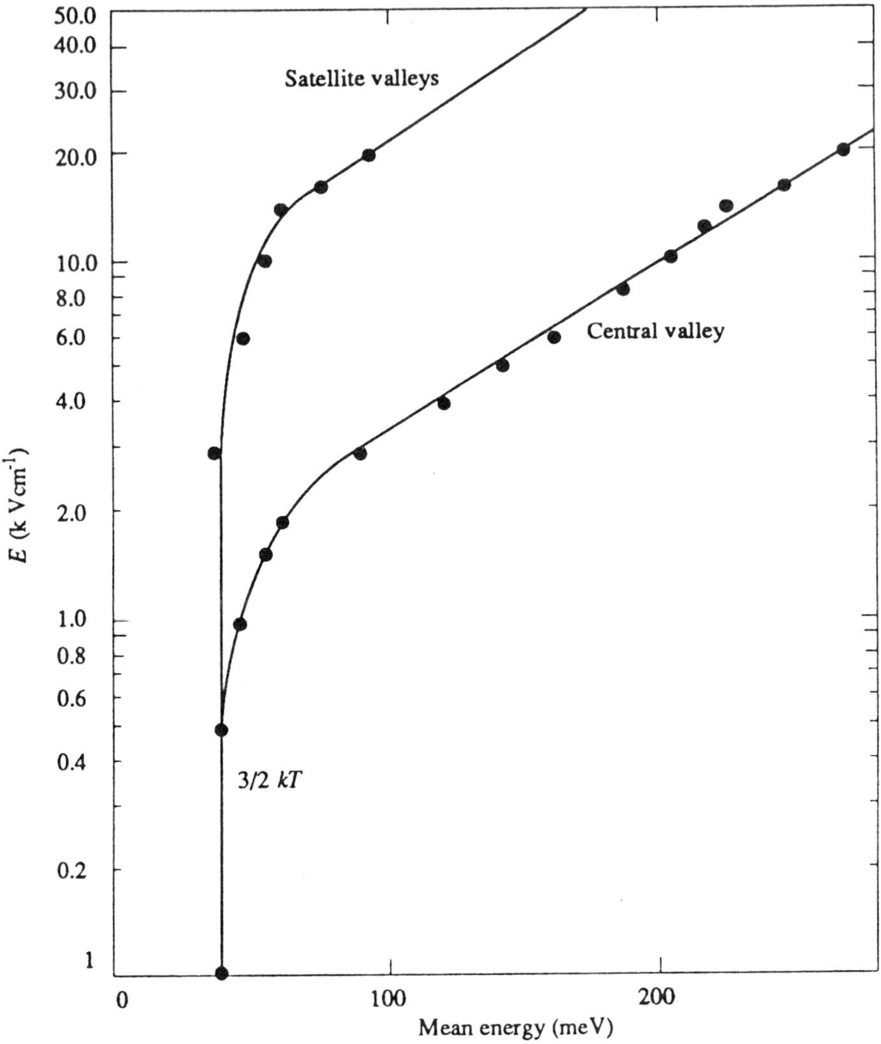

FIGURE 2

This method is adequate if E does not vary too rapidly as a function of $\mathbf{r}$ in the transistor. The average velocity, $\mathbf{v}$, is used to give the mobility, μ, for each valley.

The average $\mathscr{E}$, interpreted as $3kT/2$, gives the effective temperature, T, for each valley, The average time $\tau(E)$ for scattering from one valley to the other also comes out of the calculation.

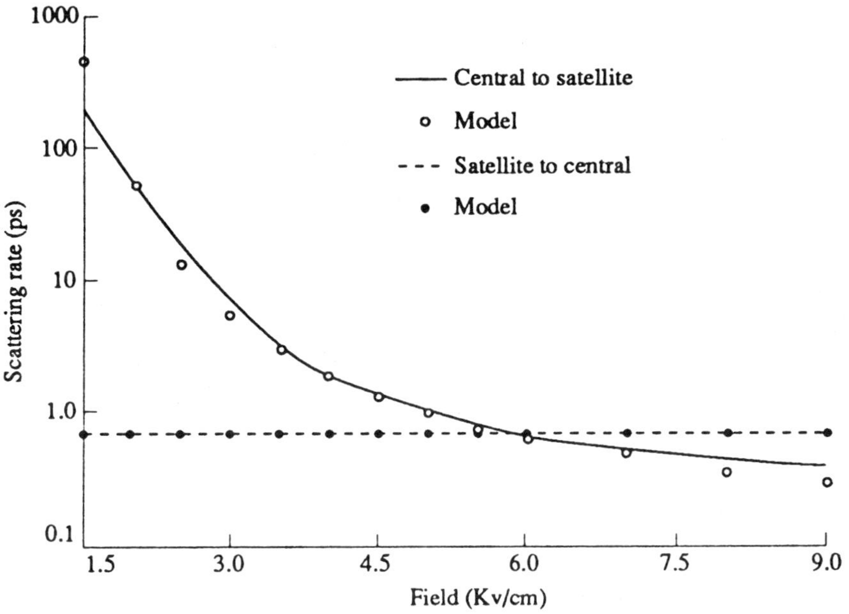

FIGURE 3

To obtain PDEs for GaAs, note that the f in each valley is simple, although the overall f is complicated. A continuity equation PDE may be found by taking the zeroth moment of f. This is adequate if the distribution in each valley is simple enough. In the satellite valley, we use

$$f_s(\mathscr{E}) = \frac{1}{1 + e^{(\mathscr{E} - q\phi + q\phi_s - \mathscr{E}_f)/(kT_s)}}.$$

In the central valley, for $\mathscr{E} < \mathscr{E}_{cs}$, we use the same kind of expression:

$$f_c(\mathscr{E}) = \frac{1}{1 + e^{(\mathscr{E} - q\phi + q\phi_c - \mathscr{E}_f)/(kT_c)}}.$$

For $\mathscr{E} > \mathscr{E}_{cs}$, we assume all the electrons have scattered out of the valley, and $f_c(\mathscr{E}) = 0$.

We define incomplete Fermi-Dirac integrals,

$$\mathscr{G}_{\frac{1}{2}}(\mathscr{E}, \mathscr{E}_{\max}, T) = \int_0^{\mathscr{E}_{\max}} \frac{1}{1 + e^{(z-\mathscr{E})/(kT)}}\, dz.$$

Then define the electron densities to be n_s and n_c:

$$n_s = N_s \mathscr{G}_{\frac{1}{2}}(q\phi - q\phi_s - \mathscr{E}_s, \infty, T_s),$$
$$n_c = N_c \mathscr{G}_{\frac{1}{2}}(q\phi - q\phi_c - \mathscr{E}_c, \mathscr{E}_{cs}, T_c).$$

Finally, we have the basic semiconductor equations for GaAs:

$$\nabla \cdot (\kappa \nabla \phi) = -\frac{q}{\varepsilon_0}(N - n_c - n_s),$$

$$\nabla \cdot (\mu_c n_c \nabla \phi_c) = \frac{\partial n_c}{\partial t} + S,$$

$$\nabla \cdot (\mu_s n_s \nabla \phi_s) = \frac{\partial n_s}{\partial t} - S,$$

where the scattering rate S is given by:

$$S = \frac{n_c}{\tau_{cs}(E)} - \frac{n_s}{\tau_{sc}(E)}.$$

5. Solving the PDEs. This is done using the B2DE package [**10**]. It solves systems of nonlinear elliptic PDEs, of the form

$$\nabla \cdot (a_i \nabla u_i) = f_i$$

with a and f nonlinear functions of $x, y, \mathbf{u}, \partial\mathbf{u}/\partial x$, and $\partial\mathbf{u}/\partial y$.

A typical geometry is shown below. (The vertical dimension is larger than shown in Figure 4.)

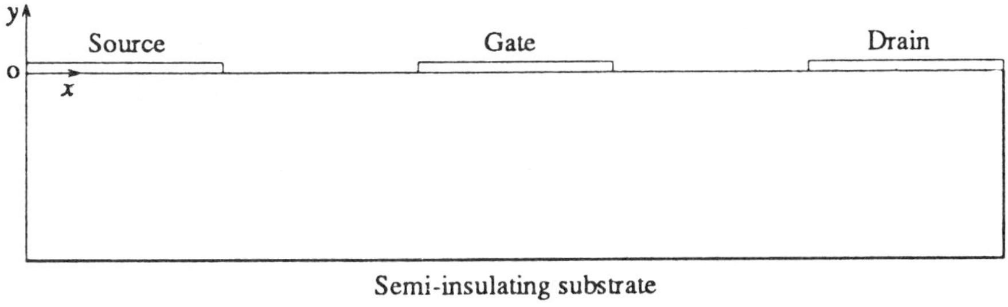

FIGURE 4

B2DE uses a nonuniform triangulation with linear finite elements, and an adaptive mesh. Following is a typical initial mesh.

Also shown are contour plots of ϕ.

6. Conclusions. Splitting of the scattering processes is critical in a two-band hydrodynamic model. When the two-band model is used it is possible to obtain reasonable agreement with experiment without introducing adjustable parameters.

Currents in the two-band model are much larger than the currents would be in a one-band model which used an average mobility; the larger currents are required for the calculation to fit experiment.

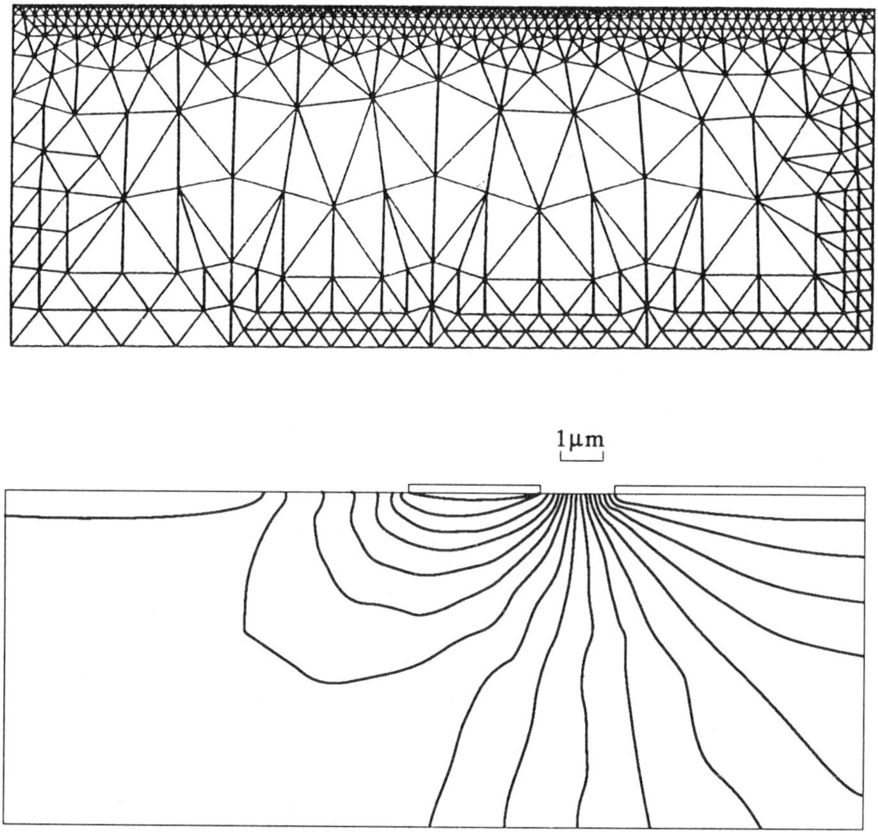

FIGURE 5

The top surface boundary condition should be a mixed boundary condition to allow the effects of the surface charge to be included in the model.

Field line termination below the back channel interface is important. This can be achieved by allowing the volume of the substrate included in the model calculation to be as deep as the top surface dimensions of the transistor.

Coupling of fields in the source-gate and gate-drain gaps is important; the geometry of the top surface metal contacts must be accurately modelled.

All of the effects listed above cause changes in the current flow which are much larger than the errors in the present model. Getting any one of these effects wrong will cause a 50% error in the calculated current even when the correct equations are used in the model.

BIBLIOGRAPHY

1. W. van Roosbroeck, *Theory of the flow of electrons and holes in Germanium and other semiconductors*, Bell Sys. Tech. J. **24** (1950), 560–607.

2. S. M. Sze, *Physics of semiconductor devices*, second edition, Wiley, New York, 1981.

3. See other papers in the current volume.

4. J. L. Blue and C. L. Wilson, *Two-dimensional analysis of short-channel MOSFETs using general-purpose interactive PDE software*, IEEE Trans. Electron Devices **ED-30** (1983), 1056–1070, published jointly in SIAM J. Sci. Statist. Comput. **4** (1983), 462–484.

5. E. M. Butturla, P. E. Cottrell, B. M. Grossman, and K. A. Salzburg, *Finite-element analysis of semiconductor devices: The FIELDAY Program*, IBM J. Res. Develop. **25** (1981), 218–231.

6. W. Fichtner, D. J. Rose, and R. E. Bank, *Semiconductor device simulation*, SIAM J. Sci. Statist. Comput. **4** (1983), 391–415.

7. C. L. Wilson, P. Roitman, and J. L. Blue, *High-accuracy physical modeling of submicrometer MOSFETs*, IEEE Trans. Electron Devices **ED-32** (1985), 1246–1258.

8. C. L. Wilson, *Hydrodynamic carrier transport in semiconductors with multiple band minima*, IEEE Trans. Electron Devices, ED-35 (1988), 180–187.

9. A. C. Smith, J. F. Janak, and R. B. Adler, *Electronic conduction in solids*, McGraw-Hill, New York, 1967.

10. J. L. Blue, *B2DE: A program for solving systems of nonlinear elliptic partial differential equations in two dimensions*, NBSIR 86-3411, National Bureau of Standards, Gaithersburg, MD, June 1986.

NATIONAL INSTITUTE OF STANDARDS AND TECHNOLOGY
GAITHERSBURG, MARYLAND 20899

Lectures in Applied Mathematics
Volume **25** (1990)

Nonlinear Mathematical Phenomena Associated with Semiconductor Devices

MEL S. BERGER

Abstract. In this article attention is focused on the nonlinear analytic aspects of semiconductor device design from the mathematical viewpoint. Among the items discussed is Shockley's equation and bifurcation in the thyristor. Assuming a doping profile for one-dimensional *p-n* is given, we develop mathematical models via the nonlinear van Roosbroeck mobility model equations for silicon devices showing that modern analytic methods in mathematics can be used for understanding and computation for semiconductor design.

1. Introduction. In the usual electronic device Ohm's Law applies. Thus electric current, denoted by the symbol I, is proportional to the applied voltage V. However in the modern silicon semiconductor *p-n* junction in one space dimension, the electric current is proportional to an exponential function of the external voltage. In symbols:

$$I = c_1[\exp(c_2 V) - 1] \tag{1}$$

where c_1 and c_2 are constants that can be explicitly calculated from the electronic characteristics of the device. Thus even the simplest semiconductor device is a highly nonlinear one. This exponential law is generally referred to as Shockley's equation after one of the pioneers of the subject.

This basic physical fact leads to the question of formulating a quantitative explanation of this observation as well as the many other switching phenomena, latch-up phenomena and other highly nonlinear semiconductor effects in terms of mathematical structure that have predictive

---

1980 *Mathematics Subject Classification* (1985 *Revision*). Primary 35B35, 78A15.
Research partially supported by an NSF grant.

capacity. This idea complements the usual numerical procedures that have proved so useful in recent years.

2. Survey of the governing equations. Now the Shockley equation can be derived from the appropriate mobility model via the one-dimensional van Roosbroeck equations (cf. Ashcroft–Mermin, *Solid State Physics*, Holt-Rinehart, 1976). This derivation utilizes the important fact that the p-n junction has been carefully "doped" (i.e., an impurity has been methodically and precisely introduced into the semiconductor device). Thus the resulting equations for the electric current can be written in the form:

$$Au = f \tag{2}$$

where f is an inhomogeneous term representing doping, A is an appropriate differential system of operators, and u is the desired unknown electric potential. The system (2) is a mathematical abstraction of the desired governing equation which is useful for mathematical motivation in the following. (The form (2) is exactly the abstract operator that I have pursued in recent years to discover global nonlinear invertible mappings and novel bifurcation phenomena.) See Berger [**1**].

Mathematically speaking, the van Roosbroeck equations can be deduced from Maxwell's equations of electromagnetism. The usual derivation results from neglecting all magnetic effects and by assuming the current j results from electrons j_n and holes j_p. Moreover, each of these currents is assumed in the mobility model to be caused by two effects: a "drift current" assumed proportional to the electric field E and a "diffusion current" which is proportional to the number of electrons n and the number of holes p (per unit volume). Moreover, one assumes the charge density is completely determined by the ionized doping.

These assumptions lead to the system:

$$\varepsilon\Delta u + q(p-n) = f, \text{ where } \Delta \text{ is the Laplace operator,} \tag{3a}$$

$$\frac{\partial n}{\partial t} = q^{-1}(\operatorname{div} j_n), \tag{3b}$$

$$\frac{\partial p}{\partial t} = q^{-1}(\operatorname{div} j_p). \tag{3c}$$

Here $E = -\operatorname{grad} u$ with u the electric potential and f the "doping profile." The dielectric denotes $D = \varepsilon E$ with ε a constant and q is the electronic charge.

Of course these equations needed to be supplemented by boundary conditions to get a well-posed problem. Thus the term "biasing" is introduced to denote positively and negatively imposed voltage Dirichlet boundary conditions on the ohmic contact (i.e., forward or reverse biasing).

The most important observable effects in semiconductor devices arise from the coupling of charge fluctuations and current response to the electric field. Thus (3b) and (3c) are often modified by adding a quantity on their right-hand side G-R to represent the possibility of local generation-recombination effect. Thus (3b) and (3c) are then rewritten

(3b′) $$\frac{\partial n}{\partial t} = q^{-1}(\operatorname{div} j_n) + (G\text{-}R)$$

and

(3c′) $$\frac{\partial p}{\partial t} = -q^{-1}(\operatorname{div} j_p) + (G\text{-}R).$$

Moreover, the current relations for the electron and hole current densities (from the mobility model) are

$$j_n = \mu_n(nE + \nabla n),$$
$$j_p = \mu_p(pE - \nabla p).$$

Here μ_n and μ_p are the mobility constants for the electrons and holes and the "so-called" Einstein relations have been used. (The units used here are normalized by setting $q = kT = n_i = 1$.)

3. The Shockley equation. The simple p-n junction in arbitrary 1-, 2-, or 3-dimensional domains has no bifurcation and a well-defined unique I-V curve is known. For thermal equilibrium and small V this follows from the analysis of Berger [**2**]; for arbitrary voltage V we refer to studies of Mock [**3**] and Jerome [**4**]. Shockley's one-dimensional analysis is well represented in the Ashcroft–Mermin text (loc. cit.). This situation is represented by the well-known I-V curve derived from (1).

However, when the doping profile is altered, the (I-V) curve changes (see Figure 3). In fact the monotone behavior of the derivative $I'(V)$ can be substantially altered and when $I'(V)$ is negative is said to exhibit "negative resistivity."

In Figure 2, I note the fact that the doping profile is piecewise constant justifying the Sobolev space approach to this problem of I-V characteristics used in Berger [**2**].

This one-dimensional analysis depends on the validity of the depletion layer hypothesis in which the mobility model is linearized over

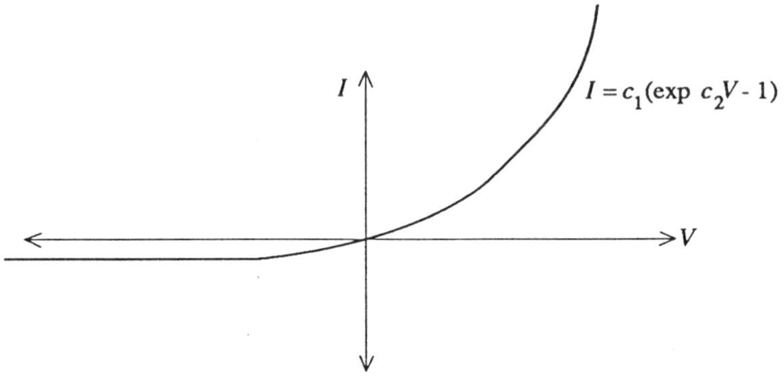

FIGURE 1. The Shockley equation for 1-D transistor

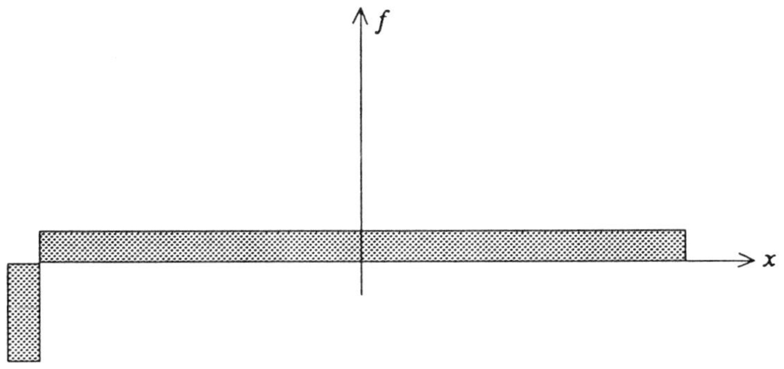

FIGURE 2. Doping profile of a 1-D, p-n junction

a certain region R of the transistor by setting $p = n = 0$. Thus on R the nonlinear Poisson equation (3a) becomes simply a nonhomogeneous linear problem which when coupled with the appropriate boundary conditions can be completely solved. The mathematical justification of this depletion layer hypothesis is an interesting outstanding problem in higher space dimensions.

4. Bifurcation phenomena in semiconductor devices. Perhaps the simplest semiconductor device exhibiting this behavior is the thyristor. The thyristor exhibits a switching phenomenon that has been explained in physical terms but whose mathematical development up to now has been unclear. In fact in terms of a J-V curve, the graph of voltage versus current exhibits negative resistivity as in Figure 3. The mathematical explanation of the switching process is as a bifurcation phenomenon.

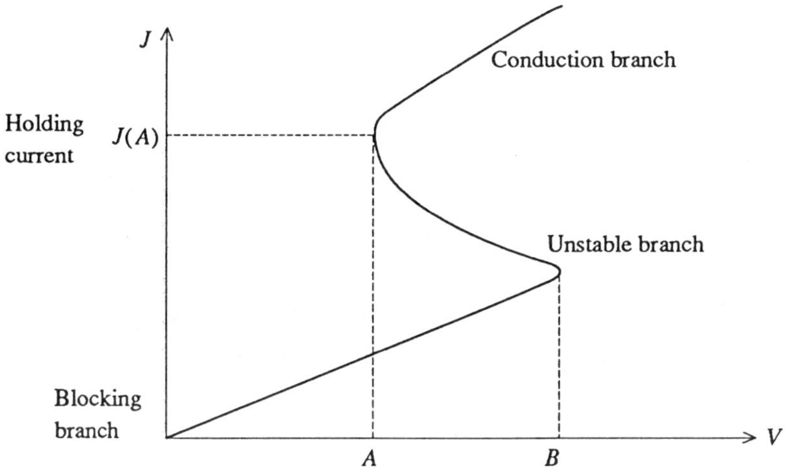

FIGURE 3. J-V curve for the thyristor

The bifurcation equations are a consequence of the mathematical structures inherent in the equation and in the boundary conditions. The important parameter involved is the magnitude of the doping profile (see Figure 4).

The important issue mathematically is the description of the behavior at the bifurcation point. The nonconvexity of the J-V graph can be made more precise by an analysis of the bifurcation problem posed in mathematical terms. This analysis is carried out in the Appendix at the end of the paper and proves the existence of bifurcation points on the J-V curve. Each bifurcation point is a fold and this leads to a justification of the turning points A and B on the V-axis.

The doping profile generating the J-V curve of Figure 3 is drawn in Figure 4. It consists of dividing the x-axis into four segments and alternating the sign of the doping profile as indicated in the figure. This doping profile coupled with the 1-D steady state van Roosbroeck equations generates the J-V curve of Figure 3.

5. The mathematics of bifurcation phenomena. For any physical system described by a system of the form (2) we can regard the differential operations and boundary conditions A as a mapping between two function spaces X and Y. Generally one studies the linearization of A, $A'(u)$, at a function u simply by writing the Taylor series

$$A(u + \varepsilon y) = A(u) + \varepsilon A'(u)y + O(|\varepsilon|^2). \tag{3}$$

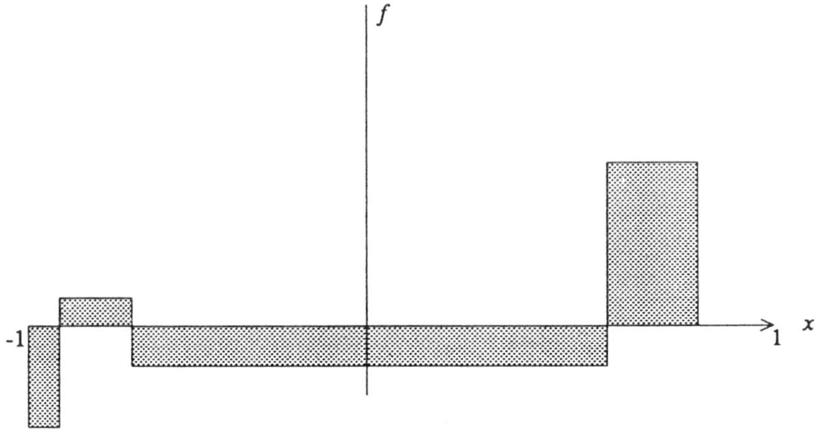

FIGURE 4. Doping profile for the thyristor

Here y is another function, and ε is a real number. In the finite-dimensional case $A'(u)$ is just the Jacobian matrix evaluated at the vector u and A is simply a vector of functions. For partial differential operators, $A'(u)$ is the linearized operator evaluated at u. For steady state problems as in the form (3) describing the thyristor we describe the onset of bifurcation by finding the functions u at which the linear operator $A'(u)$ is not invertible. Using the Fredholm alternative this means that the equation

$$A'(u)y = 0$$

supplemented by boundary conditions has a nonzero solution y. If the equation $A(v) = f$ has *nonunique solutions* for v near u and f near $A(u)$ we call u a bifurcation point.

If $A'(u)$ is always invertible no bifurcation exists and for computational purposes the Newton method or a modification of it suffices for computing numerical results.

If bifurcation does occur at u, i.e., u is a bifurcation point, there is a classification of different kinds of behavior based on the structure of the differential operator A. The simplest case is called a fold and can be characterized by the fact that near the bifurcation point u (say) all linear approximations to the behavior of A fail, but (after a change of coordinates) A can be rewritten near u in the form $(t, y) \to (t^2, y)$. This means that after a change of coordinates A has a parabolic shape (cf. Figure 3) locally.

In terms of the operator A the fold is easily distinguished by virtue of the following result proved by me and P. T. Church [**5**].

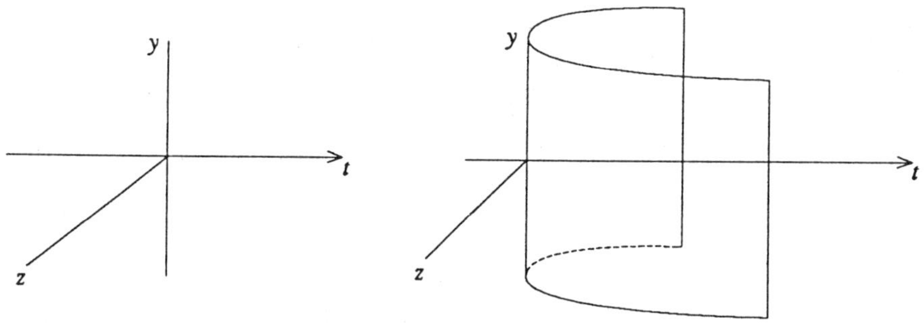

FIGURE 5. The fold $(t, y, z) \to (t^2, y, -z)$

THEOREM. *Let $A: X \to Y$ be a C' Fredholm operator of index zero acting between two real Hilbert spaces. Then a singular point x is a fold if and only if*

(i) $\dim \operatorname{Ker} A'(x) = 0$,

(ii) $(A''(x)(e, e), e^*) \neq 0$ *for nonzero* $e \in \operatorname{Ker} A'(x)$ *and nonzero* $e^* \in \operatorname{Ker}[A'(u)]^*$.

The *switching behavior* of the thyristor can be easily explained via the J-V curve. When the voltage V, the biasing, is switched on and is small the thyristor has small J as shown in the "blocking branch" before A in Figure 3. To get the thyristor into the conduction branch with high J-current without altering V, one utilizes the fact that thc "middle branch" joining the blocking and conduction branch is unstable. Thus introducing a sufficiently large perturbative gate current to the system in the blocking state causes the system to jump "over" the unstable branch into the conducting state. Conversely once in the conduction state the system can jump to the blocking state by reducing V below the holding current voltage A (as in Figure 3).

We conclude by noting that currently bifurcation phenomena are a widespread problem as in the latch-up phenomena. I predict this area to be an important research direction in the future. (See the footnote at the end of the Appendix and Pinto's article.)

Appendix: Bifurcation folds in thyristors. In a recent Ph.D. thesis, H. Steinrück carried through an analysis of the 1-D time-independent van Roosbroeck equations to get bifurcation in thyristors analytically in terms of folds. In this appendix I briefly review this work.

First the "zero space charge" approximation is used (as justified in Markowich [**6**]), i.e., $\varepsilon = 0$ is used in (3a), (3′b), (3′c) and by a linear change of coordinates the problem is reduced to the single second-order nonlinear differential equation

$$\sigma'' - \frac{|f|J}{\sigma^2}\sigma' - 2R(\sigma) = 0.$$

Here

$$\sigma = n + p, \qquad J = j_n + j_p,$$
$$n = \tfrac{1}{2}(\sigma + |f|), \qquad p = \tfrac{1}{2}(\sigma - |f|),$$

and $R(n(\sigma), p(\sigma)) = R(\sigma)$ is the Shockley–Reed–Hall recombination term.

A one-dimensional interval $[-1, 1]$ with four appropriate *p-n* junctions, with piecewise flat doping profiles (as in Figure 4) at $x_1, x_2, x_3, x_4 \in [-1, 1]$ are introduced. In the vicinity of the point $(B, J(B))$ on the *J-V* curve it is shown that new coordinates can be introduced identifying B as a bifurcation point of fold type, as described in Section 5. For the elaborate details of this analysis I refer to Steinrück's thesis [7].

Bibliography

1. M. S. Berger, *Nonlinearity and functional analysis*, Academic Press, New York, 1977.

2. ———, *Analytic aspects of VLSI*, IEEE Trans. on Electron Devices **30** (1983), 1181–1183.

3. M. S. Mock, *Analysis of semiconductor devices*, Boole Press, Dublin, 1983.

4. J. W. Jerome, *Consistency of semiconductor modeling: An existence/stability analysis for the stationary Van Roosbroeck system*, SIAM J. Appl. Math. **45** (1985), 565–590.

5. M. S. Berger and P. T. Church, *Complete integrability of a nonlinear Dirichlet problem I*, Indiana Univ. Math. J. **28** (1979), 935–952.

6. P. A. Markowich, *The stationary semiconductor device equations*, Springer, New York, 1986.

7. H. Steinrück, *A bifurcation analysis of the 1-D semiconductor device equations*, Ph.D. thesis, Tech. Univ. Wien, 1987.

Department of Mathematics and Statistics
University of Massachusetts
Amherst, Massachusetts 01003

In Boston at a 1983 IEEE meeting on computation in semiconductors I queried the audience as to bifurcation in devices. At the end of my talk (see [2]), thirty members of the audience came to speak to me concerning this topic. It is clear that the usual Newton method used in devices and circuits leaves much to be desired.

Lectures in Applied Mathematics
Volume **25** (1990)

Modular Algorithms for Transient Semiconductor Device Simulation, Part I: Analysis of the Outer Iteration

W. M. COUGHRAN, JR. AND JOSEPH W. JEROME

Abstract. A general outer iteration, based upon linearization, is introduced at discrete time steps for the one-dimensional semiconductor device model. The iteration depends upon solving the semidiscrete device equations approximately, specifically, in such a way that the residual is of order Δt in an appropriate norm. It is shown that this maintains the order of the backward Euler method. A monitoring of the constants, including time-step requirements for solvability of the semidiscrete systems, as well as boundedness, smoothness and invertibility for the maps defining the Newton approximations, is carried out. An invariant-region principle provides an important theoretical basis, and a novel proof is provided. An essential component of the theory, providing the interface with the sequel to this paper, is that the Newton iterations are small in number, typically one or two, and may be realized as approximate Newton iterations. Continuation is employed as the time-stepping bridge.

1. Introduction. In part I of this two-part series we describe a modular development for a time-stepping algorithm, based upon the backward Euler method and associated linearization, as applied to the one-dimensional semiconductor device model. We provide a detailed mathematical analysis of the solvability of the semidiscrete equations and the associated invariant region; of the residual error tolerance in solving the semidiscrete systems approximately so as to maintain first-order convergence; of an approximate Newton method, typically involving

1980 *Mathematics Subject Classification* (1985 *Revision*). Primary 65M15, 82A70; Secondary 65N40, 35K45, 47H17.

Research supported by National Science Foundation grant DMS-8420192.

one or two iterations, serving as the outer iteration which defines the approximation strategy; and of the continuation procedure, linking the output of the previous time step to the starting iterate of the current time step. Part II of the series is concerned with a specific inner iteration, i.e., the fully discrete algorithm. In decoupling the outer and inner iterations in this way, we have established a framework in Part I for any Newton-based algorithm, and have demonstrated how the a priori estimates lead to the control and estimation of fundamental constants required for a convergence analysis.

The devices studied here are one-dimensional, and the dependent variables selected are the electrostatic potential and the carrier concentrations. The model is described more fully in Section 2. The mobility coefficients are selected to have a form similar to that used in computations for silicon devices so as to ensure the physically essential property of saturation; the Einstein relations are assumed only so that physically realistic representation of diffusion is possible. For the dependent variables selected here, the Einstein relations provide no equation simplification. The model assumes Shockley–Read–Hall recombination. As described here, the model is a special case of the spatially multidimensional model considered in [**14**] by Jerome; in particular, we deduce that the initial/boundary-value problem possesses a unique solution, globally in time. Uniqueness need not persist for the semidiscrete solutions without additional time-step restrictions, however, so that the linearized time-stepping must have an inherent tracking mechanism; this is provided by the continuation.

In this paper, we do not analyze second-order time-stepping methods. We note, however, that one such method, possessing the property of L-stability, has been introduced in [**1**] by Bank et al. Issues of simulation and implementation are discussed at length in Part II. The ideas of Part I are developed fully in Sections 3 and 4. In Section 5 we provide a postscript, in which the paper is coalesced and a linkage to Part II is established.

2. The transient semiconductor equations. Our interest here is in a transient initial/boundary-value problem, in one spatial dimension, that models the behavior of a simple semiconductor device; hence, the appropriate space-time domain is $\overline{\Omega} \times [0, T_0]$ where $\Omega = (a, b) \subset \mathbf{R}$ and T_0 is the final time of interest. The transient semiconductor equations, which hold for $(x, t) \in \Omega \times (0, T_0]$, are given by

$$-\nabla \cdot (\varepsilon \nabla u) + q(n - p - N) = 0, \tag{1}$$

(2) $$q\frac{\partial n}{\partial t} - \nabla \cdot J_n = -qR_n,$$

(3) $$q\frac{\partial p}{\partial t} + \nabla \cdot J_p = -qR_p,$$

where

ε is the dielectric constant;

q is the electronic charge;

$u(x,t)$ is the electrostatic potential so that the electric field $E = -\nabla u$;

$n(x,t)$ and $p(x,t)$ are the electron and hole carrier concentrations, respectively;

$N(x)$ is the net impurity (doping) concentration;

$J_n(u,n)$ and $J_p(u,p)$ are the electron and hole current densities, respectively;

$R_n(n,p)$ and $R_p(n,p)$ are the carrier recombination-generation rates, respectively.

Note that (1) is the stationary Maxwell equation governing the electrostatic potential u while (2) and (3) are typical continuity equations governing n and p, respectively. In addition, we assume that appropriate initial data $n_0(x)$, $p_0(x)$ and boundary data $\overline{u}(x,t), \overline{n}(x), \overline{p}(x)$ are given.

We write the current densities in the traditional drift-diffusion form (cf. Sze [**19**, p. 50])

(4) $$J_n = -q\mu_n n \nabla u + qD_n \nabla n,$$

(5) $$J_p = -q\mu_p p \nabla u - qD_p \nabla p,$$

where μ_n and μ_p are the field-dependent mobilities and D_n and D_p are the electron and hole diffusion coefficients, respectively.

One simple mobility model that includes velocity saturation has the form

$$\mu(\nabla u) = \mu_0 \left[1 + \left(\frac{\mu_0|\nabla u|}{v_{\text{sat}}}\right)^{\gamma}\right]^{-1/\gamma},$$

where μ_0 and v_{sat} are the low-field mobility value and temperature-dependent saturation velocity, respectively (cf. Caughey and Thomas [**6**] and Thornber [**20**]); usually, γ is taken to be 2 for electrons and 1 for holes. To avoid problems with smoothness, we will assume $\gamma \equiv 2$

for both carriers and write the mobility functions as

$$\mu_n = \mu_{0n}\left[1 + \left(\frac{\mu_{0n}|\nabla u|}{v_{sn}}\right)^2\right]^{-1/2}, \tag{6}$$

$$\mu_p = \mu_{0p}\left[1 + \left(\frac{\mu_{0p}|\nabla u|}{v_{sp}}\right)^2\right]^{-1/2}. \tag{7}$$

Of course, there are other possibilities for the mobilities, which are less smooth or are more complicated functions of the dependent variables. It is important to note, however, that the mobility models are largely phenomenological expressions, which attempt to incorporate a number of experimentally observed phenomena.

We supplement (4) and (5) by the Einstein relations

$$D_n = \frac{kT}{q}\mu_n, \qquad D_p = \frac{kT}{q}\mu_p, \tag{8}$$

as in the Van Roosbroeck model [**21**]. Here, k and T are Boltzmann's constant and the temperature, respectively. To simplify matters, we select the natural units so that

$$\frac{kT}{q} \equiv 1; \tag{9}$$

in fact, we scale all of the equations in a fashion similar to that of DeMari [**8**], [**7**], except in our treatment of ε (see also the book by Markowich [**16**]).

Finally, we make use of the Shockley–Read–Hall recombination term [**18**], [**10**], [**17**]

$$R_n = R_p = \frac{np - 1}{\tau_p(n+1) + \tau_n(p+1)} = \frac{np-1}{d} = R, \tag{10}$$

where τ_n and τ_p are electron and hole lifetimes, respectively. Equation (10) has this simple form because the carrier concentrations n and p are measured (scaled to be) in so-called effective intrinsic carrier concentration units. Our framework can be extended to include three-particle interactions, such as Auger recombination-generation, but we will not discuss these issues here.

3. Solvability and approximate solution of the semidiscrete system. We begin by deriving a solvability result for the problem (1)–(10), which is based upon an invariant-region principle; since the complete verification of the solvability is quite detailed, we have presented the structure of the proof in §3.1 and the details of the proof in §3.2. The discussion here is related to earlier work described by Bank et al. [**2**] and Jerome [**13**]. A convergence result, where the residual is controlled in L^2, is presented in §3.3, while the corresponding H^{-1} result is presented in §3.4.

3.1. *The solvability result.* We shall find it convenient to expand the mobility terms in (2) and (3) by the product rule, substitute the second derivative terms via the potential equation (1), and make use of the Einstein relations (8) and the scaling (9). Thus, throughout much of §3.1, we shall write the semidiscrete system via a fully implicit time discretization as

$$-\varepsilon\nabla^2 u_k + n_k - p_k = N, \tag{11}$$

$$\frac{n_k - n_{k-1}}{\Delta t_k} - \nabla\cdot(\mu_n\nabla n_k) + \mu_n\nabla u_k\nabla n_k + U_{n,k} = 0, \tag{12}$$

$$\frac{p_k - p_{k-1}}{\Delta t_k} - \nabla\cdot(\mu_p\nabla p_k) - \mu_p\nabla u_k\nabla p_k + U_{p,k} = 0, \tag{13}$$

where

$$U_{n,k} = R_k + \varepsilon^{-1}(\mu_n'\nabla u_k + \mu_n)n_k(n_k - p_k - N), \tag{14}$$

$$U_{p,k} = R_k - \varepsilon^{-1}(\mu_p'\nabla u_k + \mu_p)p_k(n_k - p_k - N), \tag{15}$$

and μ_* and μ_*' are evaluated at ∇u_k.

The invariant-region principle referred to above represents a slight weakening of the usual such principle, since the region is permitted to increase linearly (not exponentially) with time and includes the spatially dependent doping. Specifically, we define a number

$$\lambda_0 = \max(\|\overline{n}\|_{L^\infty}, \|\overline{p}\|_{L^\infty}, \|n_0\|_{L^\infty}, \|p_0\|_{L^\infty}), \tag{16}$$

and *functions* $n_k^{\max}$ and $p_k^{\max}$, via the relations,

$$n_0^{\max} = \lambda_0 + \mathrm{Step}(N^+), \tag{17}$$

$$p_0^{\max} = \lambda_0 + \mathrm{Step}(N^-), \tag{18}$$

$$n_k^{\max} - n_{k-1}^{\max} = p_k^{\max} - p_{k-1}^{\max} = 4\rho\Delta t_k, \quad \text{for } 1 \le k \le L, \tag{19}$$

where $\mathrm{Step}(N^-) = \mathrm{Step}((-N)^+)$ and

$$\rho = (\tau_p + \tau_n)^{-1}, \tag{20}$$

$$\mathrm{Step}(N^+) = \begin{cases} \sup N & \text{if } N > 0, \\ 0 & \text{otherwise.} \end{cases} \tag{21}$$

It follows that, if

$$n^{\max} = \lambda_0 + \sup N^+ + 4\rho T_0, \qquad p^{\max} = \lambda_0 + \sup N^- + 4\rho T_0, \tag{22}$$

then $n^{\max} \geq n_k^{\max}$ and $p^{\max} \geq p_k^{\max}$. As before, $\overline{n}$ and $\overline{p}$ are linear extensions of the boundary data, $N = N^+ - N^-$, and T_0 is the maximum time.

We are now prepared for the first result.

THEOREM 1. *Under hypotheses on the time step (cf.* (37), (62), *and* (69) *below), there is a solution of the Dirichlet boundary-value problem* (11)–(13) *with boundary values*

$$(23)\quad u_k(x_e) = \overline{u}(x_e, t_k),\ n_k(x_e) = \overline{n}(x_e),\ p_k(x_e) = \overline{p}(x_e)\ \textit{for}\ x_e = a, b.$$

The solution triple satisfies an invariant-region principle for the carrier concentrations and a generalized maximum principle for the potential:

$$(24)\quad 0 \leq n_k \leq n_k^{\max} \leq n^{\max},$$

$$(25)\quad 0 \leq p_k \leq p_k^{\max} \leq p^{\max},$$

$$(26)\quad |u_k| \leq \|\overline{u}(\cdot, t_k)\|_{L^\infty} + \varepsilon^{-1}(e^{b-a} - 1)(n^{\max} + p^{\max}) \overset{\text{def}}{=} u_k^{\max}.$$

The solution is unique under additional conditions on the time step, described in §3.2, *and the smoothness is determined by the doping function* N; *the components are minimally in* $W^{2,\infty}$, *however. Finally, if the recombination term,* R_k, *is not included in the model, then* $\rho = 0$ *may be selected in the definition of* $n^{\max}$ *and* $p^{\max}$, *and the invariant region is independent of* T_0, *so long as* $\overline{u}$ *is bounded.*

PROOF OUTLINE. For maximum clarity, we shall outline the steps of the proof here prior to amplification in §3.2. The approach is inductive, and assumes the systems are well-defined for $j < k$, and satisfy (24), (25), and (26).

(I) Define a map T from a closed bounded convex subset K of $L^2(\Omega) \times L^2(\Omega)$ into K as follows. For

$$(27)\qquad K \overset{\text{def}}{=} \{[\tilde{n}, \tilde{p}] : 0 \leq \tilde{n} \leq n_k^{\max}, 0 \leq \tilde{p} \leq p_k^{\max}\},$$

where the inequalities are understood pointwise almost everywhere, define $u[\tilde{n}, \tilde{p}]$, with $[\tilde{n}, \tilde{p}] \in K$, as the unique solution of

$$(28)\qquad -\varepsilon\nabla^2 u + \tilde{n} - \tilde{p} = N, \quad u(a) = \overline{u}(a, t_k), \quad u(b) = \overline{u}(b, t_k).$$

With u specified by (28), set $T[\tilde{n}, \tilde{p}] = [n, p]$, where n and p are specified by the uncoupled equations

$$(29)\quad F(n) = \frac{n - n_{k-1}}{\Delta t_k} - \nabla \cdot (\mu_n \nabla n) + \mu_n \nabla u \nabla n + \widetilde{U}_n(\cdot, n, \tilde{p}) = 0,$$

$$n(a) = \overline{n}(a), \qquad n(b) = \overline{n}(b),$$

$$(30)\quad G(p) = \frac{p - p_{k-1}}{\Delta t_k} - \nabla \cdot (\mu_p \nabla p) - \mu_p \nabla u \nabla p + \widetilde{U}_p(\cdot, \tilde{n}, p) = 0,$$

$$p(a) = \overline{p}(a), \qquad p(b) = \overline{p}(b).$$

Here $\widetilde{U}_n$ and $\widetilde{U}_p$ are represented by

$$(31)\quad \widetilde{U}_n(\cdot, n, \tilde{p}) = \widetilde{R}(\cdot, n, \tilde{p}) + \varepsilon^{-1}[f(\cdot, n) - n(\tilde{p} + N)](\mu_n' \nabla u + \mu_n),$$

$$(32)\quad \widetilde{U}_p(\cdot, \tilde{n}, p) = \widetilde{R}(\cdot, \tilde{n}, p) + \varepsilon^{-1}[p(N - \tilde{n}) + g(\cdot, p)](\mu_p' \nabla u + \mu_p).$$

We shall define the functions $\widetilde{R}, f$, and g and the motivation for altering $U_{n,k}$ and $U_{p,k}$. Note that there is no a priori reason why n and p must have restricted range prior to the derivation of the invariant region principle. In order to facilitate the derivation, we introduce the function $\widetilde{R}(\cdot, n, p)$, which, for each fixed $x \in [a, b]$, agrees with $R(n, p)$ for $0 \le n \le n_k^{\max}(x)$, $0 \le p \le p_k^{\max}(x)$, but is extended outside this rectangle, to the horizontal and vertical strips uniquely intersecting in the rectangle, in such a way that the partial derivatives with respect to n and p are constant. The functions f and g have the property that, for each fixed $x \in [a, b]$, $(\partial f/\partial n)(x, n)$ and $(\partial g/\partial p)(x, p)$ agree with $2n$ and $2p$ on $[0, n_k^{\max}(x)]$ and $[0, p_k^{\max}(x)]$, respectively, and are continuously extended to have constant values outside this range.

(II) Show that T is well defined. This process involves the following:

(1) the derivation of a priori estimates for u and ∇u;

(2) the existence of solutions for (29) and (30) and derivation of corresponding a priori estimates;

(3) the verification that $[n, p] \in K$.

(III) Show that T has a fixed point, $[n, p] = [n_k, p_k]$, which coincides with the solution of (11)–(13) and (23). □

3.2. *Verification of solvability.* We provide the details of (II 1, 2, 3) and (III) in the proof of Theorem 1.

(II 1) A weak $H^2(\Omega)$ solution of (28), which is the "maximal" regularity (in fact, $W^{2,\infty}$ holds), given $\tilde{n}$ and $\tilde{p}$, satisfies the usual maximum principle

$$(33)\qquad \|u\|_{L^\infty} \le \|\overline{u}(\cdot, t_k)\|_{L^\infty} + \varepsilon^{-1}(e^{b-a} - 1)\|\tilde{n} - \tilde{p} - N\|_{L^\infty},$$

as can be seen by using the proof of Theorem 3.7 of Gilbarg and Trudinger **[9]**, and an argument by contradiction replacing the local second-derivative behavior at an extreme point with the weak formulation and corresponding first-derivative behavior. Alternatively, one could employ a smoothing argument, and take limits of the classical result. Inequality (26) follows immediately from (33). The estimates,

$$\|\nabla u\|_{L^\infty} \le C\|u\|_{H^2} \le C' u_k^{\max}, \tag{34}$$

follow from a one-dimensional Sobolev embedding, and from a standard isomorphism, respectively (cf. [9, Theorem 8.12] for the latter), as well as from the representation of $u_k^{\max}$, given by the right-hand side of (26).

(II 2) The existence of a solution of (29) can be shown by Newton's method, as we now show. The linearized equation about fixed $n \in H^2(\Omega)$ becomes, for image element $-F(n)$,

$$\begin{aligned} -F(n) = {} & \frac{\delta n}{\Delta t_k} - \nabla \cdot (\mu_n \nabla \delta n) + \mu_n \nabla u \nabla \delta n + \frac{\partial \widetilde{R}}{\partial n}(\cdot, n, \tilde{p})\delta n \\ & + \frac{1}{\varepsilon}\left(\frac{\partial f}{\partial n} - \tilde{p} - N\right)(\mu_n' \nabla u + \mu_n)\delta n, \end{aligned} \tag{35}$$

$$\delta n(a) = \delta n(b) = 0. \tag{36}$$

By direct calculation, $(\partial \widetilde{R}/\partial n)(\cdot, n, \tilde{p}) \ge 0$ and $\varepsilon^{-1}(\partial f/\partial n)(\mu_n' \nabla u + \mu_n) \ge 0$. Thus, for

$$\Delta t_k \le \varepsilon / \{2 \max(\mu_{0n}, \mu_{0p})[\max(p^{\max}, n^{\max}) + \|N\|_{L^\infty}]\}, \tag{37}$$

the coefficient of δn is nonnegative; the same condition guarantees the analogous statement for δp in the linearized hole equation. Under this format, the structural hypotheses (8.5), (8.6), and (8.8) of **[9]** are satisfied, so that the boundary-value problem (35) and (36) possesses a solution (cf. [**9**, Theorem 8.3]). An estimate for the inversion operator associated with (35) and (36) of the form,

$$\|\delta n\|_{H^2} \le C\|F(n)\|_{L^2}, \tag{38}$$

holds, with C independent of n. This follows from [**9**, Theorem 8.12] after a straight-forward preliminary estimate of $\|\delta n\|_{L^2}$ in terms of $\|F(n)\|_{L^2}$.

In addition to the uniform inversion property, (38), the map F, defining (29), has a Lipschitz continuous derivative:

$$\|F'(n_1) - F'(n_2)\|_{H^2, L^2} \le M\|n_1 - n_2\|_{H^2}, \tag{39}$$

where the constant M does not depend on n_1 or n_2. Indeed, a much stronger result holds as we shall now show. We have, for $\phi \in H^2$,

$$\begin{aligned} &\|[F'(n_1) - F'(n_2)]\phi\|_{L^2} \\ (40) \qquad &\leq \left\| \frac{\partial \widetilde{R}}{\partial n}(\cdot, n_1, \tilde{p}) - \frac{\partial \widetilde{R}}{\partial n}(\cdot, n_2, \tilde{p}) \right\|_{L^2} \|\phi\|_{H^1} \\ &\quad + \frac{1}{\varepsilon} \left\| \left(\frac{\partial f}{\partial n}(\cdot, n_1) - \frac{\partial f}{\partial n}(\cdot, n_2) \right) (\mu_n' \nabla u + \mu_n) \right\|_{L^2} \|\phi\|_{H^1} \end{aligned}$$

where we have used a standard ring property (cf. Kato [**15**]). By direct computation we have

$$(41) \qquad \left\| \left(\frac{\partial f}{\partial n}(\cdot, n_1) - \frac{\partial f}{\partial n}(\cdot, n_2) \right) (\mu_n' \nabla u + \mu_n) \right\|_{L^2} \leq 4\mu_{0n} \|n_1 - n_2\|_{L^2},$$

while, if $n_1 \geq 0$, $n_2 \geq 0$, we have,

$$\begin{aligned} &\left| \frac{\partial \widetilde{R}}{\partial n}(\cdot, n_1, \tilde{p}) - \frac{\partial \widetilde{R}}{\partial n}(\cdot, n_2, \tilde{p}) \right| \\ &\quad = \left| \frac{(\tilde{p}+1)(\tilde{p}\tau_n + \tau_p)}{[\tau_p(n_1+1) + \tau_n(\tilde{p}+1)]^2} - \frac{(\tilde{p}+1)(\tilde{p}\tau_n + \tau_p)}{[\tau_p(n_2+1) + \tau_n(\tilde{p}+1)]^2} \right| \\ &\quad \leq 2(\tilde{p}+1) \left| \frac{\tau_p(n_2 - n_1)}{[\tau_p(n_1+1) + \tau_n(\tilde{p}+1)][\tau_p(n_2+1) + \tau_n(\tilde{p}+1)]} \right| \\ &\quad \leq \left(\frac{2(\tilde{p}+1)\tau_p}{(\tau_p + \tau_n)^2} \right) |n_2 - n_1|, \end{aligned}$$

which makes use of the elementary identity,

$$\frac{1}{a^2} - \frac{1}{b^2} = \left(\frac{1}{a} - \frac{1}{b} \right) \left(\frac{1}{a} + \frac{1}{b} \right).$$

This final inequality also holds on the set where $n_1 < 0$ or $n_2 < 0$, by the way R was redefined in this region. It follows that

$$(42) \qquad \left\| \frac{\partial \widetilde{R}}{\partial n}(\cdot, n_1, \tilde{p}) - \frac{\partial \widetilde{R}}{\partial n}(\cdot, n_2, \tilde{p}) \right\|_{L^2} \leq \left(\frac{2(p^{\max}+1)\tau_p}{(\tau_p + \tau_n)^2} \right) \|n_1 - n_2\|_{L^2}$$

so that (39) follows a fortiori from (40), (41), and (42).

Now the existence result for (29) can be deduced from the existence of a homotopy solution set, as developed by the second author in [**11**], applied to the map $F(\lambda, n)$, which folds F by multiplying the term $\widetilde{U}_n$ in (29) by λ, where λ is a homotopy parameter, $0 \leq \lambda \leq 1$. The existence result for $\lambda = 0$ follows from [**9**, Theorem 8.3], and we accordingly begin the homotopy solution set at the solution of the linear equation,

$F(0,n)=0$. Although the result of [**11**, Theorem 4.1] appears to be local, in relation to the radius r of the H^2 ball B_r, on which F is defined, as compared to the length of the homotopy interval, in fact, we may select r as large as required. This is permissible since the estimates (38) and (39) readily transfer to the homotopy map, in a manner independent of r. This completes (II 2), since the boundary-value problem (30) is handled analogously, with no additional restriction on the time step.

(II 3) The choice of the functions $p_k^{\max}$ and $n_k^{\max}$ ensures the validity of the lower bounds in the inequalities

$$\widetilde{U}_n(\cdot,n,\tilde{p}) \leq |n|(\sigma + 2\varepsilon^{-1}\mu_{0n}p^{\max}), \qquad n \leq 0, \tag{43}$$

$$\widetilde{U}_n(x,n,\tilde{p}) \geq -\rho - \mu_{0n}\varepsilon^{-1}n\|N\|_{L^\infty}, \qquad n \geq n_k^{\max}(x), \tag{44}$$

$$\widetilde{U}_p(\cdot,\tilde{n},p) \leq |p|(\sigma + 2\varepsilon^{-1}\mu_{0p}n^{\max}), \qquad p \leq 0, \tag{45}$$

$$\widetilde{U}_p(x,\tilde{n},p) \geq -\rho - \mu_{0p}\varepsilon^{-1}p\|N\|_{L^\infty}, \qquad p \geq p_k^{\max}(x), \tag{46}$$

where

$$\sigma = 2\varepsilon^{-1}\max(\mu_{0n},\mu_{0p})\|N\|_{L^\infty}. \tag{47}$$

We proceed now to verify the inequalities,

$$0 \leq n \leq n_k^{\max}, \qquad 0 \leq p \leq p_k^{\max}. \tag{48}$$

Our verification of (48) will be based upon a fully discrete approximation, satisfying inequality (48), except possibly on an asymptotically thin set. Passage to the limit will yield the result. Initially, we assume that $\tilde{p}$ and $\tilde{n}$ are continuous. The equally spaced mesh points are denoted x_j and $x_{j+1/2} \overset{\text{def}}{=} (x_j + x_{j+1})/2$. Now we apply the box method to that part of the differential equation (29), exclusive of the convective term, i.e., we integrate terms in (29) over fixed subintervals $(x_{j-1/2}, x_{j+1/2})$, $1 \leq j \leq M$, with the following identifications:

$$\int_{x_{j-1/2}}^{x_{j+1/2}} -\nabla\cdot(\mu_n\nabla n)\,dx \approx \mu_n|_{x_{j-1/2}}\frac{n_j^h - n_{j-1}^h}{h} - \mu_n|_{x_{j+1/2}}\frac{n_{j+1}^h - n_j^h}{h}, \tag{49}$$

$$\int_{x_{j-1/2}}^{x_{j+1/2}} \frac{n - n_{k-1}}{\Delta t_k}\,dx \approx \frac{h}{\Delta t_k}[n_j^h - n_{k-1}(x_j)], \tag{50}$$

$$\int_{x_{j-1/2}}^{x_{j+1/2}} \widetilde{U}_n(\cdot,n,\tilde{p})\,dx \approx h\widetilde{U}_n(x_j,n_j^h,\tilde{p}(x_j)). \tag{51}$$

Here, h denotes the uniform grid spacing and n_j^h the "unknowns". The convective term, however, prior to "integration," is discretized by the following upwinding scheme:

$$\mu_n \nabla u \nabla n(x_j) \hookrightarrow (\mu_n \nabla u)(x_j)(n_j^h - n_{j-1}^h)/h, \tag{52}$$

if $(\mu_n \nabla u)(x_j) \geq 0$, while

$$\mu_n \nabla u \nabla n(x_j) \hookrightarrow (\mu_n \nabla u)(x_j)(n_{j+1}^h - n_j^h)/h, \tag{53}$$

if $(\mu_n \nabla u)(x_j) < 0$. The important properties of this discretization are now summarized:

(i) Let the discrete system be written

$$(I + L^h)n^h = n_{k-1} + B - \Delta t_k \widetilde{U}_n, \tag{54}$$

where n_{k-1} and $\widetilde{U}_n$ are column vectors, with components determined from (49), (50), and (51), and where B involves the boundary terms. Thus, if $\nabla u(x_1) > 0$,

$$0 \leq b_1 = \alpha \mu_n(\nabla u(x_{1/2}))\overline{n}(a) + \beta(\mu_n \nabla u)(x_1)\overline{n}(a)$$

is the first component of B, and subsequent components b_j satisfy $b_j = 0$, $2 \leq j \leq M - 1$, until

$$b_M = \alpha \mu_n(\nabla u(x_{M+1/2}))\overline{n}(b) + \beta(\mu_n \nabla n)(x_M)\overline{n}(b) \geq 0,$$

if $\nabla u(x_M) \leq 0$. The second terms in b_1 and b_M are zero if $\nabla u(x_1) \leq 0$, or $\nabla u(x_M) \geq 0$, respectively. Here we have used the notation $\alpha = \Delta t_k / h^2$ and $\beta = \Delta t_k / h$.

(ii) L^h has the property that its diagonal components are positive, and its off-diagonal nonpositive. By the choice of discretization (cf. (49)–(53)), it follows that L^h is weakly diagonally dominant; in fact, the upwinding has been selected to preserve the weak diagonal dominance of the box method applied to $-\nabla(\mu_n \nabla n)$. It follows (cf. **[22]**) that $\theta I + L^h$ is a diagonally dominant M-matrix for every $\theta > 0$, and, in particular, that $\theta I + L^h$ is invertible, with $(\theta I + L^h)^{-1} \geq 0$. It is also true that $\|(I + L^h)^{-1}\|_{l^\infty, l^\infty} \leq 1$, a fact which is less familiar, and therefore proved in Lemma 2, at the conclusion of (II.3). In fact, a more general result is proved.

In the light of properties (i) and (ii), we establish the inequality $n^h \geq 0$ and, up to a set Ω^h, $|\Omega^h| \to 0$ as $h \to 0$, $n^h \leq n_k^{\max}$. The desired inequality of (48) is achieved in the limit, as $h \to 0$, as we later indicate. The set Ω^h is concentrated about the finite number of points where N changes sign. A remark is in order, which will be used in both cases, i.e., if D is a nonnegative diagonal matrix, and

$$(\theta I + D + L^h)y = w, \tag{55}$$

$$(\theta I + L^h)z = w, \tag{56}$$

where $w \geq 0$ is assumed, as well as the property that $\theta I + L^h$ is a diagonally dominant M-matrix, then

$$0 \leq y \leq z. \tag{57}$$

If D is simply diagonal, then the lower bound holds when $\theta I + D + L^h$ is a diagonally dominant M-matrix. The lower bound is well known (see Berman and Plemmons **[5]**). One verifies the upper bound in (57) by simple subtraction and inversion of (55) and (56). We now proceed to verify $n^h \geq 0$. We distinguish two cases. For j, such that $n_j^h > 0$, we have

$$-[\widetilde{U}_n(\cdot, n^h, \tilde{p})]_j \geq -\left(\frac{2\mu_{0n}}{\varepsilon}\right)\left(\frac{f(x_j, n_j^h)}{n_j^h} + \|N\|_{L^\infty}\right) n_j^h,$$

while for $n_j^h \leq 0$, (43)–(46) hold. If

$$D \overset{\text{def}}{=} \operatorname{diag}[\delta_j],$$

where

$$\delta_j \overset{\text{def}}{=} \begin{cases} 0, & n_j^h \leq 0, \\ \Delta t_k[2(\varepsilon n_j^h)^{-1}\mu_{0n} f(x_j, n_j^h) + \sigma], & n_j^h > 0, \end{cases}$$

then (54) and the above yield

$$(\theta I + D + L^h) n^h \geq n_{k-1} + B \geq 0,$$

for

$$\theta \overset{\text{def}}{=} 1 - (2\varepsilon^{-1}\mu_{0n} p^{\max} + \sigma)\Delta t_k.$$

Implicit in this display is that $\theta > 0$; however, if Δt_k satisfies (62) to follow, then $\theta > 0$ does hold. Upon use of (55) and (57), and the M-matrix property of $\theta I + L^h$, we deduce that $n^h \geq 0$. In a similar way, $p^h \geq 0$ and the time-step restriction is also implied by (62).

The proof that the jth component of n^h does not exceed $n_k^{\max}(x_j)$, except in a thin set Ω^h, begins with the vector equation

$$\begin{aligned} &(I + L^h)[n^h - n_{k-1}^{\max} + \operatorname{Step}(N^+)] \\ &\quad = (n_{k-1} - n_{k-1}^{\max}) + (B - L^h[n_{k-1}^{\max} - \operatorname{Step}(N^+)]) \\ &\quad\quad + \operatorname{Step}(N^+) - \Delta t_k \widetilde{U}_n(\cdot, n^h, \tilde{p}). \end{aligned} \tag{58}$$

For j, such that $n_j^h < n_k^{\max}(x_j)$, we have, upon omission of $\tilde{p}(x_j)$ and $N^+(x_j)$, for some $0 < c < 1$,

$$\begin{aligned} -[\widetilde{U}_n(\cdot, n^h, \tilde{p})]_j \leq -c\varepsilon^{-1}\mu_{0n}\{&[(n_j^h - n_{k-1}^{\max}(x_j)) + (n_{k-1}^{\max}(x_j) - n_k^{\max}(x_j))] \\ &\cdot (n_j^h + n_k^{\max}(x_j) + N^-(x_j)) + (f(\cdot, n_k^{\max}) - n_k^{\max} N^+)(x_j) \\ &- (n_k^{\max}\tilde{p} - n_k^{\max} N^-)(x_j)\}, \end{aligned}$$

while for j, such that $n_j^h \geq n_k^{\max}(x_j)$, we have (43)–(46). If

$$D \stackrel{\text{def}}{=} \operatorname{diag}[\delta_j],$$

where

$$\delta_j \stackrel{\text{def}}{=} \begin{cases} -\Delta t_k \varepsilon^{-1} \mu_{0n} \|N\|_{L^\infty}, & n_j^h \geq n_k^{\max}(x_j), \\ \Delta t_k \varepsilon^{-1} \mu_{0n}[c(n_j^h + n_k^{\max}(x_j) + N^-(x_j)) - \|N\|_{L^\infty}], & n_j^h < n_k^{\max}(x_j), \end{cases}$$

then we obtain from (58), and the above, the estimate, where, by (62), $\delta_j \geq -1/2$,

$$\begin{aligned} &(I + D + L^h)[n^h - n_{k-1}^{\max} + \operatorname{Step}(N^+)] \\ (59) \qquad &\leq B - L^h[n_{k-1}^{\max} - \operatorname{Step}(N^+)] + (I + D)\operatorname{Step}(N^+) + \operatorname{col}(\rho)\Delta t_k \\ &\quad + 2\varepsilon^{-1}\mu_{0n}(n_k^{\max} - n_{k-1}^{\max})(n^{\max} + \sup N^-)\Delta t_k. \end{aligned}$$

The final two terms in (59) result from distinguishing the cases $n_j^h \geq n_k^{\max}(x_j)$ and $n_j^h < n_k^{\max}(x_j)$, respectively. Note that we have also been able to employ the induction hypothesis and the inequality,

$$(f(\cdot, n_k^{\max}) - n_k^{\max}\operatorname{Step}(N^+))(x_j) \geq (n_k^{\max}\tilde{p} - n_k^{\max}\operatorname{Step}(N^-))(x_j),$$

in deriving (59) from (58). In order to treat the term involving B, we note that, by design,

$$(60) \qquad n_{k-1}^{\max} - \operatorname{Step}(N^+) = \operatorname{col}(c), \qquad c = c(k),$$

for a constant $c \geq \lambda_0$, where λ_0 is defined in (16). Now the composite difference scheme annihilates constants, in the sense that

$$L^h \operatorname{col}(c) = \widetilde{B},$$

where $\widetilde{B}$ is formed from B by replacing the boundary values by c. It follows from (19) and (60), since $\widetilde{B} \geq B$, that

$$\begin{aligned} n^h - n_{k-1}^{\max} + \operatorname{Step}(N^+) \leq Q^h \Delta t_k[I + 8\varepsilon^{-1}\mu_{0n}\Delta t_k(n^{\max} + \sup N^-)]\operatorname{col}(\rho) \\ + Q^h(I + D)\operatorname{Step}(N^+), \end{aligned}$$

where we used (16)–(21), and where $Q^h \stackrel{\text{def}}{=} (I + D + L^h)^{-1} \geq 0$. This expression can be simplified by use of Lemma 2. Indeed, we obtain

$$\begin{aligned} n^h - n_{k-1}^{\max} + \operatorname{Step}(N^+) \leq 2\Delta t_k[I + 8\varepsilon^{-1}\mu_{0n}\Delta t_k(n^{\max} + \sup N^-)]\operatorname{col}(\rho) \\ + Q^h(I + D)\operatorname{Step}(N^+). \end{aligned}$$

After some simplification, we obtain

$$(61) \qquad n^h \leq n_{k-1}^{\max} - \operatorname{Step}(N^+) + Q^h(I + D)\operatorname{Step}(N^+) + 4\Delta t_k \operatorname{col}(\rho),$$

provided

$$(62) \qquad 8\varepsilon^{-1}\Delta t_k \max(\mu_{0n}, \mu_{0p})[\max(n^{\max}, p^{\max}) + \|N\|_{L^\infty}] \leq 1.$$

Note that (62) also accommodates the hole equation analysis. Now, Lemma 2 implies that

$$\|Q^h(I+D)\,\mathrm{Step}(N^+)\|_{l^\infty} \leq \sup N^+,$$

and a direct analysis shows that, for a string of zeros in $\mathrm{Step}(N^+)$, the corresponding positions in $Q^h(I+D)\,\mathrm{Step}(N^+)$ are zero, except possibly for the positions of the first and last zeros. Thus, Ω^h is defined as the union of intervals, specified by the partition points on either side of the (finitely many) points where N changes sign. This analysis shows that, for $x_j \notin \Omega^h$,

$$n^h(x_j) \leq n_{k-1}^{\max}(x_j) + 4\rho\Delta t_k = n_k^{\max}(x_j).$$

In particular, the upper bound has been demonstrated for the discrete solutions, except on Ω^h, since a similar argument, and result, hold for p^h, with a possible enlargement of Ω^h.

If we knew that solutions of (29) and (30) were smooth, then well-known arguments associated with the convergence of the box method, augmented by a straightforward analysis of the upwinding approximation, would give the required bounds (48), and would permit $\widetilde{U}_n$ and $\widetilde{U}_p$ to be replaced by U_n and U_p in (29) and (30). Since C^3 solution regularity is sufficient for $O(h)$ convergence, we may assume that the result holds for this class. To achieve this, we could mollify the nonnegative L^2 functions $\tilde{n}$ and $\tilde{p}$ to obtain smooth nonnegative functions for which the corresponding C^3 solutions of (29) and (30) are nonnegative. However, these functions need not satisfy the upper bounds. Instead, we define, for $\delta > 0$,

$$n_\delta(x) = \begin{cases} \tilde{n}(x), & x \notin \Omega^\delta, \\ 0, & x \in \Omega^\delta, \end{cases}$$

where Ω^δ is a set of measure proportional to δ and centered about the points where N changes sign. On each of the complementary subintervals, $n_k^{\max}$ reduces to a constant upper bound, though the constant itself changes from interval to interval. The individual function components of n_δ can be mollified, with the support of the mollified components contained within the subinterval, in such a way that the bounds are satisfied, i.e., the mollified pair is in K. In this way, one obtains an L^2 convergent sequence of approximations to $\tilde{n}$ (and $\tilde{p}$). It is a standard result, and, in fact, follows from a slight generalization of the analysis of (III) to follow, that this yields L^2 convergence of the solutions of the mollified problems to the fixed solutions of (29) and (30). This completes the verification of parts (1), (2), and (3) of (II), except for Lemma 2, which we present now.

LEMMA 2. *Let L^h be the weakly diagonally dominant matrix of* (54), *with positive diagonal entries, and nonpositive off-diagonal entries. Let D be any diagonal matrix, $D = \operatorname{diag}[\delta_j]$, with $\delta_j > -1$. Then*

$$\|(I + D + L^h)^{-1}(I + D)\|_{l^\infty, l^\infty} \leq 1. \tag{63}$$

If $\delta_j \geq -1/2$, then

$$\|(I + D + L^h)^{-1}\|_{l^\infty, l^\infty} \leq 2. \tag{64}$$

PROOF. Let y and z satisfy

$$(I + D + L^h)y = (I + D)z,$$

and let j denote an index for which $|y_j| = \|y\|_{l^\infty}$. Without loss of generality we assume $y_j \geq 0$. Then,

$$(1 + \delta_j)y_j \leq (1 + \delta_j)z_j + y_j \left(-l_{jj} + \sum_{m \neq j} l_{jm} \right),$$

where we have denoted the diagonal elements of L^h by l_{jj} and the off-diagonal elements by $-l_{jm}$. Since weak diagonal dominance assures that

$$-l_{jj} + \sum_{m \neq j} l_{jm} \leq 0,$$

we conclude that

$$0 \leq y_j \leq z_j \leq \|z\|_{l^\infty}.$$

This establishes the first statement of the lemma, since a priori considerations assure that $(I + D + L^h)^{-1}$ exists. The proof of the second statement is similar. □

(III) We shall directly estimate the Lipschitz constant of $T\colon K \to K$. Thus, let

$$u_1 = u_1[\tilde{n}_1, \tilde{p}_1], \qquad u_2 = u_2[\tilde{n}_2, \tilde{p}_2]. \tag{65}$$

The first step consists of the estimate,

$$\begin{aligned} \|u_1 - u_2\|_{L^2}^2 &\leq \left(\frac{b - a}{\pi}\right)^2 \|u_1 - u_2\|_{H^1}^2 \\ &\leq \left(\frac{2(b - a)^4}{\pi^4 \varepsilon}\right) \|[\tilde{n}_1 - \tilde{n}_2, \tilde{p}_1 - \tilde{p}_2]\|_{L^2 \times L^2}^2, \end{aligned} \tag{66}$$

which follows upon subtracting the respective formulations of (28) for $[\tilde{n}_1, \tilde{p}_1]$ and $[\tilde{n}_2, \tilde{p}_2]$, using $u_1 - u_2$ as a test function in the weak version, estimating inner products in terms of sums of squares, and, finally, estimating the L^2 norm in terms of the H^1 norm. The next step uses a variant in which (29) and (30) are collapsed back so that $\nabla \cdot (\mu_n \nabla un)$ and $-\nabla \cdot (\mu_p \nabla up)$ are the relevant drift terms. In this format, we subtract the relevant equations involving n_1 and n_2 as well as p_1 and p_2 if

$$[n_1, p_1] = T[\tilde{n}_1, \tilde{p}_1], \qquad [n_2, p_2] = T[\tilde{n}_2, \tilde{p}_2]. \tag{67}$$

Using the monotonicity of the recombination terms, in each variable separately, and using $n_1 - n_2$, $p_1 - p_2$ as test functions, respectively, we obtain, for $\lambda > 0$,

(68)
$$\begin{aligned} &\|n_1 - n_2\|_{L^2}^2 + \Delta t_k \|\sqrt{\mu_n(\nabla u_1)}\nabla(n_1 - n_2)\|_{L^2}^2 \\ &\quad \le \Delta t_k \left\{ \frac{\lambda}{2} \|\nabla n_2\|_{L^\infty}^2 \|\mu_n(\nabla u_1) - \mu_n(\nabla u_2)\|_{L^2}^2 + \frac{1}{2\lambda}\|\nabla(n_1 - n_2)\|_{L^2}^2 \right. \\ &\qquad + \frac{\lambda}{2}\|n_2\|_{L^\infty}^2 \|\mu_n(\nabla u_1)\nabla u_1 - \mu_n(\nabla u_2)\nabla u_2\|_{L^2}^2 + \frac{1}{2\lambda}\|\nabla(n_1 - n_2)\|_{L^2}^2 \\ &\qquad \left. + \frac{\lambda}{2}\|\mu_n(\nabla u_1)\nabla u_1\|_{L^\infty}^2 \|n_1 - n_2\|_{L^2}^2 + \frac{1}{2\lambda}\|\nabla(n_1 - n_2)\|_{L^2}^2 \right\}. \end{aligned}$$

A similar inequality holds for the p equation. The functions $\xi \mapsto \mu_n(\xi)$ and $\xi \mapsto \xi\mu_n(\xi)$ are Lipschitz continuous, with constants μ_{0n}^2 / v_{sn} and $2\mu_{0n}$ so that, combined with the a priori estimates for $\nabla u_1, n_2$, and ∇n_2 this yields, for $\lambda = 2/\inf \mu_n(\nabla u_1)$,

$$\|n_1 - n_2\|_{L^2}^2 \le \Delta t_k \left[\left(\frac{v_{sn}^2}{\inf \mu_n} \right) \|n_1 - n_2\|_{L^2}^2 + C\|u_1 - u_2\|_{H^1}^2 \right],$$

where C does not depend on $\tilde{n}_1, \tilde{p}_1, \tilde{n}_2, \tilde{p}_2$. A similar inequality holds for $\|p_1 - p_2\|_{L^2}$. These computations show that T is Lipschitz continuous if

$$\Delta t_k < \frac{\min(\inf \mu_n, \inf \mu_p)}{\max(v_{sn}^2, v_{sp}^2)}. \tag{69}$$

Rather than make an unreasonable further restriction to guarantee that T is a strict contraction, we may apply the Schauder fixed-point theorem to T to conclude the argument. The necessary H^1 bounds are obtained from the second term on the left-hand side of (68). Uniqueness, however, is not obvious without the further restriction that T be a strict contraction.

3.3. *Approximate solvability and L^2 residual control.* A typical computing procedure involves solving the system (11)–(13) only approximately. In the next major result, we present and analyze a criterion for approximate solvability in terms of the residuals of the individual systems at successive time steps. The computed triple, $[u_k, n_k, p_k]$, need not satisfy the invariant-region property as presented in Theorem 1. Since pointwise bounds are essential to the theory, they are built into the hypothesis structure directly at the outset.

THEOREM 3. *Suppose that the triple $[u_k, n_k, p_k]$ approximately solves* (11)–(13), *i.e., the boundary conditions* (23) *are satisfied exactly and*

$$-\varepsilon\nabla^2 u_k + n_k - p_k - N = r_1, \tag{70}$$

$$\frac{n_k - n_{k-1}}{\Delta t_k} - \nabla\cdot(\mu_n\nabla n_k) + \mu_n\nabla u_k\nabla n_k + U_{n,k} = r_2, \tag{71}$$

$$\frac{p_k - p_{k-1}}{\Delta t_k} - \nabla\cdot(\mu_p\nabla p_k) - \mu_p\nabla u_k\nabla p_k + U_{p,k} = r_3, \tag{72}$$

where the residuals r_1, r_2, and r_3 satisfy

$$\|r_1\|_{L^2}^2 \le C_u(\Delta t_k)^2, \quad \|r_2\|_{L^2} \le C_n\Delta t_k, \quad \|r_3\|_{L^2} \le C_p\Delta t_k, \tag{73}$$

for $k = 1, \dots, L$. Define the errors,

$$v_k \stackrel{\text{def}}{=} u(\cdot, t_k) - u_k, \tag{74}$$

$$e_k \stackrel{\text{def}}{=} n(\cdot, t_k) - n_k, \tag{75}$$

$$q_k \stackrel{\text{def}}{=} p(\cdot, t_k) - p_k. \tag{76}$$

Suppose that pointwise bounds,

$$\|u_k\|_{L^\infty} \le c_u, \quad \|n_k\|_{L^\infty} \le c_n, \quad \|p_k\|_{L^\infty} \le c_p, \tag{77}$$

exist for the computed semidiscrete approximations, together with the lower bounds,

$$1 + n_k \ge 0, \quad 1 + p_k \ge 0. \tag{78}$$

Define

$$\Delta t \stackrel{\text{def}}{=} \max_k \Delta t_k. \tag{79}$$

If $1 - c_1\Delta t > 0$ with c_1 specified in (96) *below, then these errors converge* *with optimal order in L^2:*

$$\sup_{1\le M\le L}(\|e_M\|_{L^2}^2+\|q_M\|_{L^2}^2) + \min(\inf\mu_n,\inf\mu_p)\sum_{k=1}^{L}(\|\nabla e_k\|_{L^2}^2+\|\nabla q_k\|_{L^2}^2)\Delta t_k \le C_1(\Delta t)^2, \tag{80}$$

$$\sup_{1\le M\le L}\|\nabla v_M\|_{L^2}^2 \le C_2(\Delta t)^2, \tag{81}$$

where C_1 and C_2 are certain positive constants, given explicitly in (98) *and* (99) *below. The lower bounds,* (78), *are unnecessary if $R\equiv 0$ is selected. The result remains valid if n^2 is replaced by $(n_+)^2$ and p^2 by $(p_+)^2$ in $U_{n,k}$ and $U_{p,k}$, respectively.*

PROOF. Subtract (71) from the second equation, (2), of the device system to obtain

$$\begin{aligned}(\Delta t_k)^{-1}&(e_k-e_{k-1})-\nabla(\mu_n(\nabla u_k)\nabla e_k)\\ &+\nabla([\mu_n(\nabla u_k)-\mu_n(\nabla u(\cdot,t_k))]\nabla n(\cdot,t_k))+\mu_n(\nabla u_k)\nabla u_k\nabla e_k\\ &-[\mu_n(\nabla u_k)\nabla u_k-\mu_n(\nabla u(\cdot,t_k)\nabla u(\cdot,t_k))]\nabla n(\cdot,t_k)\\ &=-r_2+V_{n,k}+(\Delta t_k)^{-1}\int_{t_{k-1}}^{t_k}(t_{k-1}-s)\frac{\partial^2 n}{\partial t^2}(s)\,ds,\end{aligned} \tag{82}$$

where

$$V_{n,k}=-[U_n(u(\cdot,t_k),n(\cdot,t_k),p(\cdot,t_k))-U_{n,k}(u_k,n_k,p_k)]. \tag{83}$$

In the estimation to follow, we shall write

(84)

$$\begin{aligned}V_{n,k}=&[U_n(u,n,p_k)-U_n(u,n,p)]-[R(n,p_k)-R(n_k,p_k)]\\ &-\varepsilon^{-1}[(\mu_n'(\nabla u)\nabla u+\mu_n(\nabla u))-(\mu_n'(\nabla u_k)\nabla u_k+\mu_n(\nabla u_k))]\\ &\quad\cdot[n^2-p_k n-Nn]\\ &-\varepsilon^{-1}[\mu_n'(\nabla u_k)\nabla u_k+\mu_n(\nabla u_k)][(n^2-n_k^2)-p_k(n-n_k)-N(n-n_k)].\end{aligned}$$

Multiplication of (82) by $e_k \Delta t_k$, integration over Ω, and summation on $k = 1, \dots, M$ for $1 \le M \le L$ yield, after some simplification,

$$\begin{aligned}
&\frac{1}{2}\|e_M\|_{L^2}^2 + \inf \mu_n \sum_{k=1}^{M} \|\nabla e_k\|_{L^2}^2 \Delta t_k \\
&\le \Bigg\{ \left(\frac{\mu_{0n}^2}{v_{sn}}\right) \|\nabla n\|_{L^\infty} \sum_{k=1}^{M} \|\nabla v_k\|_{L^2} \|\nabla e_k\|_{L^2} + v_{sn} \sum_{k=1}^{M} \|\nabla e_k\|_{L^2} \|e_k\|_{L^2} \\
&\quad + 2\mu_{0n} \|\nabla n\|_{L^\infty} \sum_{k=1}^{M} \|\nabla v_k\|_{L^2} \|e_k\|_{L^2} + C_n \sum_{k=1}^{M} \|e_k\|_{L^2} \Delta t_k \\
&\quad + \sum_{k=1}^{M} \|e_k\|_{L^2} \|f_k\|_{L^2} + C_{1n} \sum_{k=1}^{M} \|e_k\|_{L^2}^2 \\
&\quad + C_{2n} \sum_{k=1}^{M} \|q_k\|_{L^2} \|e_k\|_{L^2} + C_{3n} \sum_{k=1}^{M} \|\nabla v_k\|_{L^2} \|e_k\|_{L^2} \Bigg\} \Delta t_k.
\end{aligned} \tag{85}$$

Here,

$$f_k = (\Delta t_k)^{-1} \int_{t_{k-1}}^{t_k} (t_{k-1} - s) \frac{\partial^2 n}{\partial t^2}(s)\, ds, \tag{86}$$

and we have noticed that $\xi \mapsto \mu_n(\xi)$, $\xi \mapsto \xi \mu_n(\xi)$, and $\xi \mapsto \xi \mu_n'(\xi)$ have respective Lipschitz constants of μ_{0n}^2 / v_{sn}, $2\mu_{0n}$, and $3\mu_{0n}^2 / v_{sn}$. Also we have used the monotonicity, in (84), of $R(\cdot, p_k)$ and of the quadratic in n. Finally, C_{1n}, C_{2n}, and C_{3n} are given here by

$$C_{1n} = \varepsilon^{-1} \mu_{0n} (c_p + \|N\|_{L^\infty}), \tag{87}$$

$$C_{2n} = \frac{\tau_p + \tau_n}{\tau_p^2} + \varepsilon^{-1} \mu_{0n} \|n\|_{L^\infty}, \tag{88}$$

$$C_{3n} = \left(\frac{4\mu_{0n}^2}{v_{sn}\varepsilon}\right) [\|n\|_{L^\infty}^2 + \|n\|_{L^\infty}(c_p + \|N\|_{L^\infty})]. \tag{89}$$

The first term defining C_{2n} embodies an upper bound for $\partial R / \partial p$, via the mean value theorem, and this was obtained by use of (78).

A similar inequality utilizing (72) and (3) holds for q_k. By use of (70) and (1) we deduce the important relation,

$$\|\nabla v_k\|_{L^2}^2 \le \left(\frac{4(b-a)^2}{\pi^2 \varepsilon}\right) (\|e_k\|_{L^2}^2 + \|q_k\|_{L^2}^2 + \|r_1\|_{L^2}^2). \tag{90}$$

Note that we have employed here an elementary version of Rayleigh's inequality, and made judicious use of the inequality,

$$\alpha\beta \le \frac{1}{2}(\lambda \alpha^2 + \lambda^{-1} \beta^2), \tag{91}$$

as applied to the estimated weak version of the subtracted equations. The term, $C_n\|e_k\|_{L^2}(\Delta t_k)^2$, is expanded as

$$C_n\|e_k\|_{L^2}(\Delta t_k)^2 \le \frac{1}{2}C_n\|e_k\|_{L^2}^2\Delta t_k + \frac{1}{2}C_n(\Delta t_k)^3, \tag{92}$$

with a similar inequality for the C_p term. The estimate,

$$\|f_k\|_{L^2}^2 \le \Delta t_k \left\|\frac{\partial^2 n}{\partial t^2}\right\|_{L^2(\Omega\times(t_{k-1},t_k))}^2, \tag{93}$$

is also used, with its counterpart involving p. The inequality (91) is employed to estimate the remaining terms. With proper choice of α and β, the choice, $\lambda = 1$, is made, except for those terms involving $\|\nabla e_k\|$, where $\beta = \|\nabla e_k\|_{L^2}$, $\lambda = 2/\inf \mu_n$ are appropriate choices as applied to the inequality (91). Similar choices hold for the companion inequality. The Gronwall inequality requires the absorption of the terms, $\|e_M\|_{L^2}^2$ and $\|q_M\|_{L^2}^2$, when arising on the right-hand side of the summed inequalities, to be transferred to the left-hand side. This imposes a restriction on Δt_k. Altogether, we have, upon adding the inequalities for e_k and q_k,

(94)

$$\begin{aligned}\|e_M\|_{L^2}^2 + \|q_M\|_{L^2}^2 &+ \min(\inf\mu_n, \inf\mu_p)\sum_{k=1}^{M}(\|\nabla e_k\|_{L^2}^2 + \|\nabla q_k\|_{L^2}^2)\Delta t_k \\ &\le (1 - c_1 \sup \Delta t_k)^{-1}\left[c_0(\Delta t)^2 + c_1\sum_{k=1}^{M-1}(\|e_k\|_{L^2}^2 + \|q_k\|_{L^2}^2)\Delta t_k\right],\end{aligned}$$

where

$$(95)\ c_0 = \kappa C_u + (C_n + C_p)T_0 + \|n_{tt}\|_{L^2(\Omega\times(0,T_0))}^2 + \|p_{tt}\|_{L^2(\Omega\times(0,T_0))}^2,$$

$$\begin{aligned}(96)\ c_1 &= \kappa + 2 + 2\max(C_{1n}, C_{1p}) + \max(C_n, C_p) + 2\max(C_{2n}, C_{2p}) \\ &\quad + \max(C_{3n}, C_{3p}) + \frac{2\max(v_{sn}^2, v_{sp}^2)}{\min(\inf\mu_n, \inf\mu_p)},\end{aligned}$$

$$\begin{aligned}(97)\ \kappa &= \left(\frac{4(b-a)^2}{\pi^2\varepsilon}\right)\Bigg\{C_{3n} + C_{3p} + 2(\|\nabla n\|_{L^\infty}^2 + \|\nabla p\|_{L^\infty}^2) \\ &\quad \cdot \left[\frac{\max(\mu_{0n}^4/v_{sn}^2, \mu_{0p}^4/v_{sp}^2)}{\min(\inf\mu_n, \inf\mu_p)} + 2\max(\mu_{0n}^2, \mu_{0p}^2)\right]\Bigg\}.\end{aligned}$$

Since the inequality (94) holds for each $M = 1, \ldots, L$, it is a consequence of the Gronwall inequality (cf. Jerome [**12**, pp. 52–54]) that

(80) and (81) hold with

$$C_1 = \frac{c_0 \exp[c_1 T_0/(1 - c_1 \Delta t)]}{1 - c_1 \Delta t}, \tag{98}$$

$$C_2 = \left(\frac{4(b-a)^2}{\pi^2 \varepsilon}\right)(C_u + C_1). \quad \square \tag{99}$$

REMARK 1. The advantage of the replacement of n^2 and p^2 by $(n_+)^2$ and $(p_+)^2$ is that it leads to uniform boundedness of approximate solutions (cf. Lemma 7).

3.4. *H^{-1} Residual control.* We shall develop a necessary parallel result for the case where the residual is estimated in H^{-1}. This has important implications for classes of algorithms, such as the transport diffusion algorithm, or those using second-order numerical methods, which lead naturally to H^{-1} residual estimation. In order to provide as efficient an estimation procedure as possible, we introduce the family of Riesz maps T_a, associated with the linear differential operators,

$$-\nabla a \nabla \colon H_0^1(\Omega) \to H^{-1}(\Omega). \tag{100}$$

T_a satisfies the relation, for $l \in H^{-1}(\Omega)$,

$$\langle l, v\rangle = (a\nabla T_a l, \nabla v)_{L^2}, \qquad v \in H_0^1(\Omega). \tag{101}$$

Here, $a \geq a_0 > 0$ is an essentially bounded function and will be identified with μ_n or μ_p. We note the fundamental relation,

$$(f, T_a f)_{L^2} \geq \inf a (f, T_e f)_{L^2} \geq (\inf a / \sup a)(f, T_a f)_{L^2}, \tag{102}$$

where $e(x) \equiv 1$, which follows from (101). Here, we have interpreted the duality pairing $\langle \cdot, \cdot \rangle$ as the L^2 pivot space inner product for $f \in L^2(\Omega)$. We choose, for the (equivalent) H^{-1} norm,

$$\|l\|_{H^{-1}} = \langle l, T_e l\rangle^{1/2}. \tag{103}$$

The maps T_a, when restricted to $L^2(\Omega)$, are positive definite, self-adjoint, compact operators; they are also (pointwise) nonnegative operators. Moreover, if g is Lipschitz continuous,

$$\|g(v) - g(w)\|_{H^{-1}} \leq \|g\|_{\mathrm{Lip}} \| |v - w| \|_{H^{-1}}, \qquad v, w \in L^2(\Omega). \tag{104}$$

In the extension of Theorem 3 to the H^{-1} case, one is compelled to make systematic use of (101)–(104) and the properties just mentioned, together with their implications, as well as one additional result, contained in (106). To describe this, let

$$c_a \geq \sup a. \tag{105}$$

Now we have, for $f \in H^1(\Omega)$ and $g \in W^{1,\infty}(\Omega)$ with $\|g\|_{L^\infty}^2 + \|\nabla g\|_{L^\infty}^2 \le \gamma^2$,

$$\|T_a^{1/2}(g\nabla f)\|_{L^2} \le c_a \|T_e^{1/2}(g\nabla f)\|_{L^2} \le C_a \|f\|_{L^2}, \tag{106}$$

as can be seen from the following computations. Since T_e is the Dirichlet solution operator for $-\nabla^2$,

$$T_e^{1/2}[g\nabla f(x)] = \sum_{m=1}^{\infty} \frac{a_m}{\sqrt{\lambda_m}} \sin\left(\frac{m\pi(x-a)}{b-a}\right),$$

where

$$a_m = \frac{2}{b-a}\int_a^b f'(s)g(s)\sin\left(\frac{m\pi(s-a)}{b-a}\right)ds$$

and $\lambda_m = [m\pi/(b-a)]^2$ denotes the mth eigenvalue of $-\nabla^2$. We have,

$$\begin{aligned}\left(\frac{b-a}{2}\right)a_m &= -\sqrt{\lambda_m}\int_a^b f(s)g(s)\cos\left(\frac{m\pi(s-a)}{b-a}\right)ds \\ &\quad - \int_a^b f(s)g'(s)\sin\left(\frac{m\pi(s-a)}{b-a}\right)ds \\ &\overset{\text{def}}{=} \left(\frac{b-a}{2}\right)(-\sqrt{\lambda_m}b_m - c_m).\end{aligned}$$

Now, if N_0 denotes the Neumann operator for $-\nabla^2$, i.e.,

$$-\nabla^2 N_0 f = f - \frac{1}{b-a}\int_\Omega f, \quad \nabla N_0 f(a) = \nabla N_0 f(b) = 0, \quad \int_\Omega N_0 f = \int_\Omega f,$$

it follows from the above that

$$N_0\left(fg - \frac{1}{b-a}\int_\Omega fg\right) = \sum_{m=1}^{\infty}\frac{b_m}{\lambda_m}\cos\left(\frac{m\pi(x-a)}{b-a}\right),$$

hence that (cf. [**12**, pp. 17, 18])

$$\begin{aligned}\|T_e^{1/2}(g\nabla f)\|_{L^2}^2 &= \frac{b-a}{2}\sum_{m=1}^{\infty}\frac{a_m^2}{\lambda_m} \\ &\le (b-a)\left(\sum_{m=1}^{\infty} b_m^2 + \sum_{m=1}^{\infty}\frac{c_m^2}{\lambda_m}\right) \\ &= 2\left\|fg - \frac{1}{b-a}\int_\Omega fg\right\|_{L^2}^2 + 2\|fg'\|_{H^{-1}}^2 \\ &\le 2\left(1 + \frac{(b-a)^2}{\pi^2}\right)\gamma^2\|f\|_{L^2}^2.\end{aligned} \tag{107}$$

Inequality (107) immediately yields (106) with $C_a = c_a\sqrt{2}[1 + (b-a)^2/\pi^2]^{1/2}\gamma$.

THEOREM 4. *Suppose the hypothesis,* (73), *is restated in terms of* H^{-1} *norms of the individual residuals:*

$$(108)\qquad \|r_1\|^2_{H^{-1}} \le C_u(\Delta t_k)^2, \quad \|r_2\|_{H^{-1}} \le C_n \Delta t_k, \quad \|r_3\|_{H^{-1}} \le C_p \Delta t_k.$$

Suppose that the pointwise bounds (77) *and* (78) *hold and that* r_1 *is uniformly pointwise bounded. Then*

(109)

$$\sup_{1\le M\le L} (\|e_M\|^2_{H^{-1}} + \|q_M\|^2_{H^{-1}}) + \sum_{k=1}^{L} (\|\nabla e_k\|^2_{L^2} + \|\nabla q_k\|^2_{L^2})\Delta t_k \le C_1(\Delta t^2),$$

$$(110)\qquad \sup_{1\le M\le L} \|v_M\|^2_{L^2} \le C_2(\Delta t)^2,$$

where C_1 *and* C_2 *are given explicitly below in* (116) *and* (117).

PROOF. Apply T_{μ_n} to (82) and compute the L^2 inner product with $e_k \Delta t_k$. We shall make use of the properties of T_a described earlier in order to estimate the resultant. The corresponding replacement for (85) is given by

$$\begin{aligned}
&\frac{1}{2}(\inf \mu_n)\|e_M\|^2_{H^{-1}} + \sum_{k=1}^{M} \|e_k\|^2_{L^2}\Delta t_k \\
&\quad\le \Bigg[C_{\mu_n}\left(\frac{\mu_{0n}^{5/2}}{v_{sn}}\right)\|\nabla n\|_{L^\infty} \sum_{k=1}^{M} \|\nabla v_k\|_{L^2}\|e_k\|_{H^{-1}} \\
&\qquad + C_{\mu_n}\sqrt{\mu_{0n}} \sum_{k=1}^{M} \|e_k\|_{L^2}\|e_k\|_{H^{-1}} \\
&\qquad + 2\mu_{0n}^2\left(\frac{b-a}{\pi}\right)\|\nabla n\|_{L^\infty} \sum_{k=1}^{M} \|\nabla v_k\|_{L^2}\|e_k\|_{H^{-1}} \\
(111)\quad &\qquad + C_n\mu_{0n} \sum_{k=1}^{M} \|e_k\|_{H^{-1}}\Delta t_k + \mu_{0n} \sum_{k=1}^{M} \|e_k\|_{H^{-1}}\|f_k\|_{H^{-1}} \\
&\qquad + C'_{1n}\mu_{0n}\left(\frac{b-a}{\pi}\right) \sum_{k=1}^{M} \|e_k\|_{L^2}\|e_k\|_{H^{-1}} \\
&\qquad + C_{2n}\mu_{0n}\left(\frac{b-a}{\pi}\right) \sum_{k=1}^{M} \|q_k\|_{L^2}\|e_k\|_{H^{-1}} \\
&\qquad + C_{3n}\mu_{0n}\left(\frac{b-a}{\pi}\right) \sum_{k=1}^{M} \|\nabla v_k\|_{L^2}\|e_k\|_{H^{-1}} \Bigg] \Delta t_k
\end{aligned}$$

with $C'_{1n} = C_{1n} + \|n\|_{L^\infty} + c_n$. The reader will note that $\gamma = 1$ in the first appearance of C_{μ_n} above, whereas $\gamma = (v_{sn}^2 + 4\mu_{0n}^2\omega^2)^{1/2}$ in the second appearance; here, ω is an upper bound for $\|\Delta u_k\|_{L^\infty}$. The replacement for (90) is

(112)
$$\|\nabla v_k\|_{H^{-1}}^2 = \|v_k\|_{L^2}^2 \leq \left(\frac{4(b-a)^2}{\pi^2\varepsilon}\right)(\|e_k\|_{H^{-1}}^2 + \|q_k\|_{H^{-1}}^2 + \|r_1\|_{H^{-1}}^2).$$

The remainder of the proof proceeds similarly to the proof of Theorem 3. The new choices of c_0 and c_1 are given by

$$(113)\quad c_0 = \kappa C_u + T_0\bar{\mu}(C_n + C_p + \|(T^{1/2}n)_{tt}\|_{L^2(\Omega\times(0,T_0))}^2 + \|(T^{1/2}p)_{tt}\|_{L^2(\Omega\times(0,T_0))}^2),$$

$$(114)\quad c_1 = \kappa + \bar{\mu}^2\left[1 + 6\max(C_{1n}'^2, C_{1p}'^2) + 6\left(\frac{b-a}{\pi}\right)^2 \max(C_{2n}^2, C_{2p}^2)\right] + \bar{\mu}[\max(C_n, C_p) + 6\sup C_\mu^2],$$

$$(115)\quad \kappa = \frac{3\pi^2\varepsilon}{2(b-a)^2}\left\{\left[\max\left(\frac{\mu_{0n}^5}{v_{sn}^2}, \frac{\mu_{0p}^5}{v_{sp}^2}\right)\sup C_\mu^2 + 4\bar{\mu}^4\left(\frac{b-a}{\pi}\right)^2\right] \cdot(\|\nabla n\|_{L^\infty}^2 + \|\nabla p\|_{L^\infty}^2) + \bar{\mu}^2\left(\frac{b-a}{\pi}\right)^2(C_{3n}^2 + C_{3p}^2)\right\}$$

with $\bar{\mu} = \max(\mu_{0n}, \mu_{0p})$ and $C'_{1p} = C_{1p} + \|p\|_{L^\infty} + c_p$. In (114) and (115), $\sup C_\mu^2$ refers to all choices of $\mu = \mu_n, \mu_p$, obtained from the semidiscrete system. The finiteness of this follows, ultimately, from the pointwise boundedness assumptions.

The new choices of C_1 and C_2 are given by

$$(116)\qquad C_1 = \frac{c_0\exp\{c_1T_0/[\min(\inf\mu_n, \inf\mu_p) - c_1\Delta t]\}}{\min(\inf\mu_n, \inf\mu_p) - c_1\Delta t},$$

$$(117)\qquad C_2 = \left(\frac{4(b-a)^2}{\pi^2\varepsilon}\right)(C_u + C_1). \quad \square$$

4. Approximate Newton methods for the semidiscrete system. We shall describe a rather general approximate Newton method, in which the approximate linear inversion is based upon an arbitrary inner iteration, arranged in such a way that the algorithm is quadratically convergent. Though the description is general, we shall aim for the development of an algorithm with a small number of outer iterations. The reader will recall from Section 3 that it is the residual of the nonlinear map, defining the semidiscrete system, which is to be controlled of order Δt_k (cf. Theorem 3). It follows that the analysis of the approximate Newton method of this section will address this property, although, typically, quadratic convergence of the residuals occurs in tandem with quadratic convergence of the approximants themselves.

The plan of the section is as follows. An operator tableau for the linearized problem is presented in §4.1, while the properties required of an approximate Newton method are presented in §4.2. A study of these properties, as they affect the semiconductor model, is presented next (§4.3 and §4.4), followed by a description of continuation from the previous discrete time in §4.5.

4.1. *The linearized problem.* This is primarily a formal subsection, in which we briefly present the operator for the linearized problem. At the conclusion, we mention considerations of operator domain and range.

Suppose the system (11)–(13) is denoted by

$$F(u_k, n_k, p_k) = 0, \tag{118}$$

and the mth linear increment, i.e., the difference between the mth and the $(m-1)$th Newton iterates, is denoted by $[\phi_k^m, \psi_k^m, \omega_k^m]$. This increment satisfies

$$\begin{bmatrix} -\varepsilon\nabla^2\star & \star & -\star \\ (F')_{21} & (F')_{22} & (\partial U_p/\partial p)\star \\ (F')_{31} & (\partial U_n/\partial n)\star & (F')_{33} \end{bmatrix} \begin{pmatrix} \phi_k^m \\ \psi_k^m \\ \omega_k^m \end{pmatrix} = -\begin{pmatrix} r_1^{m-1} \\ r_2^{m-1} \\ r_3^{m-1} \end{pmatrix} \tag{119}$$

where

(120)

$$(F')_{21} \stackrel{\text{def}}{=} -\nabla[\nabla n_k^{m-1}\mu_n'(\nabla u_k^{m-1})\nabla\star] + \nabla n_k^{m-1}\nabla u_k^{m-1}\mu_n'(\nabla u_k^{m-1})\nabla\star + \nabla n_k^{m-1}\mu_n(\nabla u_k^{m-1})\nabla\star + \frac{\partial U_n}{\partial u}\nabla\star$$

(121)

$$(F')_{22} \stackrel{\text{def}}{=} \frac{\star}{\Delta t_k} - \nabla[\mu_n(\nabla u_k^{m-1})\nabla\star] + \mu_n(\nabla u_k^{m-1})\nabla u_k^{m-1}\nabla\star + \frac{\partial U_n}{\partial n}\star$$

(122)

$$(F')_{31} \stackrel{\text{def}}{=} -\nabla[\nabla p_k^{m-1}\mu_p'(\nabla u_k^{m-1})\nabla\star] - \nabla p_k^{m-1}\nabla u_k^{m-1}\mu_p'(\nabla u_k^{m-1})\nabla\star - \nabla p_k^{m-1}\mu_p(\nabla u_k^{m-1})\nabla\star + \frac{\partial U_p}{\partial u}\nabla\star$$

(123)

$$(F')_{33} \stackrel{\text{def}}{=} \frac{\star}{\Delta t_k} - \nabla[\mu_p(\nabla u_k^{m-1})\nabla\star] - \mu_p(\nabla u_k^{m-1})\nabla u_k^{m-1}\nabla\star + \frac{\partial U_p}{\partial p}\star.$$

Moreover, r_1^{m-1}, r_2^{m-1}, and r_3^{m-1} are the residuals of $F(u_k^{m-1}, n_k^{m-1}, p_k^{m-1})$, given explicitly in (11)–(13); $[u_k^{m-1}, n_k^{m-1}, p_k^{m-1}]$ is the Newton iterate; and

$$\begin{aligned}\frac{\partial U_n}{\partial u} &\overset{\text{def}}{=} \frac{\partial U_n}{\partial u}(u_k^{m-1}, n_k^{m-1}, p_k^{m-1}) \\ &= \varepsilon^{-1}[2\mu_n'(\nabla u_k^{m-1}) + \mu_n''(\nabla u_k^{m-1})\nabla u_k^{m-1}] \\ &\quad \cdot [(n_k^{m-1})^2 - n_k^{m-1}p_k^{m-1} - n_k^{m-1}N]\end{aligned} \tag{124}$$

$$\begin{aligned}\frac{\partial U_p}{\partial u} &\overset{\text{def}}{=} \frac{\partial U_p}{\partial u}(u_k^{m-1}, n_k^{m-1}, p_k^{m-1}) \\ &= -\varepsilon^{-1}[2\mu_p'(\nabla u_k^{m-1}) + \mu_p''(\nabla u_k^{m-1})\nabla u_k^{m-1}] \\ &\quad \cdot [n_k^{m-1}p_k^{m-1} - (p_k^{m-1})^2 - p_k^{m-1}N]\end{aligned} \tag{125}$$

$$\begin{aligned}\frac{\partial U_n}{\partial n} &\overset{\text{def}}{=} \frac{\partial U_n}{\partial n}(u_k^{m-1}, n_k^{m-1}, p_k^{m-1}) \\ &= \frac{(\tau_p + p_k^{m-1}\tau_n)(p_k^{m-1}+1)}{(d_k^{m-1})^2} \\ &\quad + \frac{1}{\varepsilon}[\mu_n(\nabla u_k^{m-1}) + \mu_n'(\nabla u_k^{m-1})\nabla u_k^{m-1}](2n_k^{m-1} - p_k^{m-1} - N)\end{aligned} \tag{126}$$

$$\begin{aligned}\frac{\partial U_p}{\partial p} &\overset{\text{def}}{=} \frac{\partial U_p}{\partial p}(u_k^{m-1}, n_k^{m-1}, p_k^{m-1}) \\ &= \frac{(\tau_n + n_k^{m-1}\tau_p)(n_k^{m-1}+1)}{(d_k^{m-1})^2} \\ &\quad - \frac{1}{\varepsilon}[\mu_p(\nabla u_k^{m-1}) + \mu_p'(\nabla u_k^{m-1})\nabla u_k^{m-1}](n_k^{m-1} - 2p_k^{m-1} - N)\end{aligned} \tag{127}$$

with

$$d_k^{m-1} \overset{\text{def}}{=} \tau_p(n_k^{m-1}+1) + \tau_n(p_k^{m-1}+1). \tag{128}$$

We shall find it advantageous to consider two distinct application frameworks for the mappings F and F'. Note that (119)–(123) is a delineation of the equation,

$$F'(u_k^{m-1}, n_k^{m-1}, p_k^{m-1}) \begin{pmatrix} \phi_k^m \\ \psi_k^m \\ \omega_k^m \end{pmatrix} = -\begin{pmatrix} r_1^{m-1} \\ r_2^{m-1} \\ r_3^{m-1} \end{pmatrix}. \tag{129}$$

(Also note that (119) represents a considerable notational economy; a similar tableau was introduced by Bank et al. [**4**].) We shall permit both an L^2 and H^{-1} residual measurement framework because of the variety of applications. Thus, F will be defined from triples in the affine subspace of $H^s(\Omega)$ given by

$$X_s = [\overline{u}(\cdot, t_k), \overline{n}, \overline{p}] + Y_s, \quad s = 1, 2, \tag{130}$$

where

$$Y_s = \prod_1^3 H^s(\Omega) \cap H_0^1(\Omega), \quad s = 1, 2, \tag{131}$$

to the triple product of copies of $H^{s-2}(\Omega)$. $F'(z)$, for fixed z, is defined from Y_s to the same space. Any approximate inverse, denoted $G(z)$ in the next subsection, reverses domain and range. Finally, we shall find it convenient, and, in fact, necessary to restrict F to have local domain of definition.

4.2. *A class of approximate Newton methods.* We begin by recalling the three fundamental properties required of an approximate Newton method, based on a family of linear maps, $G(z^m)$, where $\{z^m\}$ comprises the Newton sequence, defined by

$$z^m - z^{m-1} \stackrel{\text{def}}{=} -G(z^{m-1})F(z^{m-1}), \tag{132}$$

and where $z, F(z) = 0$, is sought as $z = \lim_{m\to\infty} z^m$. The three properties are:

$$\|G(z^{m-1})F(z^{m-1})\| \leq M_1\|F(z^{m-1})\|, \tag{133}$$

$$\|[F'(z^{m-1})G(z^{m-1}) - I]F(z^{m-1})\| \leq M_2\|F(z^{m-1})\|^2, \tag{134}$$

$$\|F'(x) - F'(y)\| \leq 2M_3\|x - y\|, \tag{135}$$

together with a mechanism insuring that successive iterates lie within the domain of definition of F. Here, the constants M_1, M_2, and M_3 are independent of the elements in the domain of definition of the respective maps. This framework is reminiscent of that described by Bank and Rose [**3**].

In order to estimate $\|F(z^m)\|$, we write

$$F(z^m) = -[F'(z^{m-1})G(z^{m-1}) - I]F(z^{m-1}) + R(z^{m-1}, z^m) \tag{136}$$

where

$$R(z^{m-1}, z^m) = F(z^m) - F(z^{m-1}) - F'(z^{m-1})(z^m - z^{m-1}). \tag{137}$$

Notice that the first term in (136) may be dominated by $M_2\|F(z^{m-1})\|^2$, via (134), and the second term by

$$\|R(z^{m-1}, z^m)\| = \left\| \int_0^1 [F'(z^{m-1} + s(z^m - z^{m-1})) - F'(z^{m-1}](z^m - z^{m-1})\, ds \right\|$$

on use of a standard Taylor expansion. From these remarks we have the estimate

$$\|F(z^m)\| \leq (M_1^2 M_3 + M_2)\|F(z^{m-1})\|^2. \tag{138}$$

We have proved the bulk of the following lemma.

LEMMA 5. *Let F be a map defined on a closed ball B_r in an affine subspace $x_0 + X_0$, contained in a Banach space X, with Fréchet derivative*

$$F'(z)\colon X_0 \mapsto W, \qquad z \in B_r.$$

Here W is a Banach space containing the range of F. Suppose that $G(z)\colon W \mapsto X_0$ is defined for each $z \in B_r$, and that F, F', and G satisfy (133), (134), *and* (135). *If z^0 satisfies*

$$M_1\|F(z^0)\| \leq (1 - \alpha)r, \tag{139}$$

where $0 \leq \alpha < 1$ is such that $z^0 \in B_{\alpha r}$, then $z^1 \in B_r$ and the residual $F(z^1)$ satisfies

$$\|F(z^1)\| \leq (M_1^2 M_3 + M_2)\|F(z^0)\|^2. \tag{140}$$

REMARK 2. The hypothesis (135) can be weakened from $x, y \in B_r$ to x, y on the line segment between z^{m-1} and z^m. This remark is especially applicable to hypotheses (170) and (171) to follow, which need only hold along such line segments, and, in fact, can only be expected to hold in such a local fashion as induced by numerical approximation schemes.

4.3. *Estimation of the constants.* We shall estimate the constant M_3 of the previous section, and determine an estimate for M_1 in the special case $G = (F')^{-1}$. In the following section we shall show how this provides constant estimates for general G. As a preliminary comment, we recall again that the carrier approximations, unlike the solutions, can assume negative values. Some mechanism in any computing process must be found to ensure that the recombination term is well-defined. In practice, one might wish to redefine this term for negative values of

n and p, when the term appears explicitly (as opposed to implicitly). In any event, we shall assume that the denominators d_k^{m-1} of the recombination term are bounded away from zero, independently of m and k:

$$d_k^{m-1} \geq d > 0. \tag{141}$$

Hypothesis (141) is an intrinsic hypothesis associated with the model. We now estimate $(F')^{-1}$.

PROPOSITION 6. *Let F be the map of* (118), *defined on a ball of radius r_0 in X_s. The inversion map of* (129) *satisfies, for $m \leq m_0$,*

$$\|[F'(u_k^{m-1}, n_k^{m-1}, p_k^{m-1})]^{-1} r^{m-1}\|_{H^s} \leq M\|r^{m-1}\|_{H^{s-2}}, \tag{142}$$

where M is a constant depending only on r_0, m_0, and a bound for $\|\nabla^2 u_k^0\|_{L^\infty}$. For $s = 2$, there are no further unspecified hypotheses, whereas for $s = 1$ the a posteriori hypotheses (157) *and* (158) *of Lemma 7 are assumed to hold. Alternatively, in lieu of* (157) *and* (158), *the result holds if*

$$\|\nabla n_k^m\|_{L^\infty} \leq C_{\|\mathscr{P}\|}\|\nabla n_k^m\|_{L^2}, \qquad m \leq m_0, \tag{143}$$

with $C_{\|\mathscr{P}\|}^2 \|\mathscr{P}\| \to 0$ as $\|\mathscr{P}\| \to 0$.

PROOF. We begin with the case $s = 2$. From the tableau (119)–(123) we obtain the estimates

$$\begin{aligned}(\Delta t_k)^{-1}\|\psi_k^m\|_{L^2}^2 &+ \frac{1}{2}\inf \mu_n \|\nabla \psi_k^m\|_{L^2}^2 \\ &\leq \frac{1}{2}\|r_2^{m-1}\|_{L^2}^2 + \left(\frac{A^2}{\inf \mu_n}\right)\|\nabla \phi_k^m\|_{L^2}^2 \\ &\quad + \left(\frac{5}{4} + \frac{v_{sn}^2}{\inf \mu_n} + \sup\left|\frac{\partial U_n}{\partial n}\right|\right)\|\psi_k^m\|_{L^2}^2 + \frac{1}{2}\sup\left|\frac{\partial U_p}{\partial p}\right|^2 \|\omega_k^m\|_{L^2}^2\end{aligned} \tag{144}$$

and

$$\|\nabla \phi_k^m\|_{L^2}^2 \leq \left(\frac{4(b-a)^2}{\pi^2 \varepsilon}\right)(\|\psi_k^m\|_{L^2}^2 + \|\omega_k^m\|_{L^2}^2 + \|r_1^{m-1}\|_{L^2}^2). \tag{145}$$

Here,

$$A = \mu_{0n}\left[\left(\frac{\mu_{0n}}{v_{sn}}\right)\sup|\nabla n_k^{m-1}| + \sqrt{\inf \mu_n}\left(2\sup|\nabla n_k^{m-1}| + \sup\left|\frac{\partial U_n}{\partial u}\right|\right)\right]. \tag{146}$$

In conjunction with the estimate for ω_k^m, we obtain, from (144) and (145),

$$\|\psi_k^m\|_{L^2}^2 + \|\omega_k^m\|_{L^2}^2 \leq c\|r^{m-1}\|_{L^2}^2, \tag{147}$$

where c is the positive number defined by

(148)

$$c^{-1} = \left(\frac{1}{2} + \frac{4(b-a)^2}{\pi^2 \varepsilon}\right)^{-1} \left\{1 - \|\mathscr{P}\| \left[\frac{4A^2(b-a)^2}{\pi^2 \varepsilon \min(\inf \mu_n, \inf \mu_p)} + \max\left(\sup\left|\frac{\partial U_n}{\partial n}\right|^2, \sup\left|\frac{\partial U_p}{\partial p}\right|^2\right) + \frac{9}{4} + \frac{\max(v_{sn}^2, v_{sp}^2)}{\min(\inf \mu_n, \inf \mu_p)}\right]\right\}.$$

Clearly, $A \le A_0$ if the H^2 norm is restricted as specified.

The estimate (147) implies

$$\|\nabla^2 \phi_k^m\|_{L^2}^2 \le 4\varepsilon^{-2}(\|r_1^{m-1}\|_{L^2}^2 + c\|r^{m-1}\|_{L^2}^2), \tag{149}$$

which, in conjunction with

$$\|\phi_k^m\|_{L^2}^2 \le 4\varepsilon^{-1}\left(\frac{b-a}{\pi}\right)^4 (\|r_1^{m-1}\|_{L^2}^2 + c\|r^{m-1}\|_{L^2}^2), \tag{150}$$

yields an H^2 estimate for ϕ_k^m. The bootstrapping can now proceed. One views the second equation defined by (119)–(123) as having the form

$$(\Delta t_k)^{-1}\psi_k^m - \mu_n \nabla^2 \psi_k^m - \mu_n' \nabla^2 u_k^{m-1} \nabla \psi_k^m + \mu_n \nabla u_k^{m-1} \nabla \psi_k^m = -r_2^{m-1} + q_k^{m-1}, \tag{151}$$

where

$$\|q_k^{m-1}\|_{L^2} \le C\|r^{m-1}\|_{L^2}. \tag{152}$$

The latter inequality follows from (149) and (150) and the tableau display (119)–(123). Note that the constant C depends only on r_0.

If (151) is multiplied by ψ_k^m, with the second and third terms collapsed into divergence form, then integration by parts yields the estimate,

$$\|\nabla \psi_k^m\|_{L^2}^2 \le (1 + C^2)(\inf \mu_n)^{-1}\|r^{m-1}\|_{L^2}^2, \tag{153}$$

provided

$$1 - \|\mathscr{P}\| - \frac{1}{2}\|\mathscr{P}\| \max(v_{sn}^2, v_{sp}^2)/\min(\inf \mu_n, \inf \mu_p) > 0. \tag{154}$$

Here we have used (152). For the estimate (153), there is a corresponding estimate for $\nabla \omega_k^m$.

If (151) is multiplied by $-\nabla^2 \psi_k^m$ and the resultant integrated over Ω, one has

$$(\Delta t_k)^{-1}\|\nabla \psi_k^m\|_{L^2}^2 + \frac{1}{2}(\inf \mu_n)\|\nabla^2 \psi_k^m\|_{L^2}^2 \le 2(\inf \mu_n)^{-1}(\mu_{0n}^2 \|\nabla^2 u_k^{m-1}\|_{L^\infty}^2 + v_{sn}^2)\|\nabla \psi_k^m\|_{L^2}^2 + 2(1 + C^2)\|r^{m-1}\|_{L^2}^2. \tag{155}$$

The conjunction of (153), (155), and the inequality

$$\|\nabla^2 u_k^{m-1}\|_{L^\infty} \leq \|\nabla^2 u_k^0\|_{L^\infty} + (m_0 - 1) \sup_m \|\phi_k^m\|_{L^\infty} \tag{156}$$

for $m \leq m_0$, leads to (142) in the case $s = 2$. Note that (149) and (150) yield a bound for the supremum in (156).

With one major exception, the proof for $s = 1$ proceeds as above, where the Riesz maps are employed as in §3.4. The exception relates to uniform gradient bounds for n_k^m and p_k^m, which, in the case $s = 2$, are assured by the action of the mapping F on a bounded set in H^2. The lemma to follow provides these bounds, but specifies additional residual control necessary to achieve the bounds. This explains the residual hypothesis in the case $s = 1$. The alternative hypothesis, which bounds the gradient pointwise according to (143), does not require Lemma 7, and enters in (148) so that $c > 0$. In the interest of brevity, we omit the remaining details. □

LEMMA 7. *Suppose that, for $u_k = u_k^m$, $n_k = n_k^m$, and $p_k = p_k^m$, the L^2-norm of the residual associated with* (70)–(72), *denoted $r_k = [r_{1,k}, r_{2,k}, r_{3,k}]$, is square summable bounded, independently of the partition $\mathscr{P}$ and the level of the Newton iteration*:

$$\sum_k \|r_k\|_{L^2}^2 \leq \alpha \tag{157}$$

and that the residuals are uniformly bounded in L^∞:

$$\|r_k\|_{L^\infty} \leq \beta. \tag{158}$$

Suppose that the quadratic terms n^2 and p^2 are replaced by $(n_+)^2$ and $(p_+)^2$ in $U_{n,k}$ and $U_{p,k}$, respectively. Then the solution triples are bounded in $\prod_1^3 H^2$:

$$\|[u_k, n_k, p_k]\|_{H^2} \leq C. \tag{159}$$

PROOF. Although it appears natural to use the system (70)–(72) to derive these estimates, the nonlinear diffusion coefficients present technical obstacles to this approach. Somewhat surprisingly, it is advantageous to use a perturbation argument employing the solution. The latter does, however, require preliminary L^∞ estimates, for which it is possible to use (70)–(72). In order to obtain a format conducive to integration by parts, we subtract the boundary values. More precisely, set

$$\nu_k \stackrel{\text{def}}{=} n_k - \overline{n}, \tag{160}$$

so that

(161)
$$\frac{\nu_k - \nu_{k-1}}{\Delta t_k} - \nabla(\mu_n \nabla \nu_k) = \nabla(\mu_n \nabla \overline{n}) - \mu_n \nabla u_k \nabla \nu_k - \mu_n \nabla u_k \nabla \overline{n} - U_{n,k} + r_{2,k}$$

is the equivalent version of (71). If (161) is multiplied by $\nu_k^{\gamma-1}$, for γ an even integer, then an application of the inequality

$$\alpha\beta \le \frac{\alpha^\gamma}{\gamma} + \frac{\beta^{1-1/\gamma}}{1-1/\gamma}, \qquad \alpha \ge 0,\ \beta \ge 0,\ \gamma > 0 \tag{162}$$

yields, after integration over Ω,

$$\begin{aligned}
\int_\Omega |\nu_k|^\gamma &+ (\gamma-1)\Delta t_k \int_\Omega \mu_n |\nabla \nu_k|^2 \nu_k^{\gamma-2} \\
&\le v_{sn} \Delta t_k \int_\Omega |\nabla \nu_k| |\nu_k|^{\gamma-1} \\
&\quad + \int_\Omega |(\nabla(\mu_n \nabla \overline{n}) - \mu_n \nabla u_k \nabla \overline{n} - \widehat{U}_{n,k} + r_{2,k})\Delta t_k + \nu_{k-1}| |\nu_k|^{\gamma-1} \\
&\le \frac{v_{sn}^2 \Delta t_k}{2 \inf \mu_n (\gamma-1)} \int_\Omega |\nu_k|^\gamma + \frac{\gamma-1}{2} \Delta t_k \int_\Omega \mu_n |\nabla \nu_k|^2 \nu_k^{\gamma-2} \\
&\quad + \frac{1}{\gamma} \int_\Omega |(\nabla(\mu_n \nabla \overline{n}) - \mu_n \nabla u_k \nabla \overline{n} - \widehat{U}_{n,k} + r_{2,k})\Delta t_k + \nu_{k-1}|^\gamma \\
&\quad + \frac{\gamma-1}{\gamma} \int_\Omega |\nu_k|^\gamma .
\end{aligned}$$

Here we have recognized that $U_{n,k}(u_k, \cdot, p_k)$, where $\cdot$ stands for $\nu_k + \overline{n}$, can be written as a sum of two terms, one of which is $\widehat{U}_{n,k}$, and the other of which has the same sign as ν_k, $\widehat{U}_{n,k}$ has the property that it is of linear growth in its final two arguments (cf. (164)); the other summand, when multiplied by $\nu_k^{\gamma-1}$, may be neglected. A parallel statement holds for $U_{p,k}$.

Multiplication by γ and extraction of γth roots leads to (cf. (154))

(163)
$$\begin{aligned}
2^{-1/\gamma} \|\nu_k\|_{L^\gamma} \le \|\nu_{k-1}\|_{L^\gamma} &+ c\Delta t_k \left[\left(\frac{\mu_{0n}^2}{v_{sn}} \right) \|\nabla^2 u_k\|_{L^\gamma} + v_{sn} |\Omega|^{1/\gamma} \right] \\
&+ \Delta t_k [\|\widehat{U}_{n,k}\|_{L^\gamma} + \|r_{2,k}\|_{L^\gamma}],
\end{aligned}$$

where $c = |\nabla \overline{n}|$ is a constant. We may now let $\gamma \to \infty$. There is a parallel expression for $\overline{\omega}_k \stackrel{\text{def}}{=} p_k - \overline{p}$. The linear growth of $\widehat{U}_{n,k}$ and $\widehat{U}_{p,k}$ and Equation (70) lead to expressions of the form

$$\|\widehat{U}_{n,k}\|_{L^\infty} + \|\widehat{U}_{p,k}\|_{L^\infty} + \|\nabla^2 u_k\|_{L^\infty} \le c_1(\|n_k\|_{L^\infty} + \|p_k\|_{L^\infty}) + c_0, \tag{164}$$

where c_0 does not depend on k. The Gronwall inequality now leads to an exponential bound for $\|\nu_k\|_{L^\infty}$ and $\|\overline{\omega}_k\|_{L^\infty}$.

We may now proceed to second-derivative estimates. Multiplication of (82) by $-\nabla^2 e_k$ gives the algebraic relation

$$
\begin{aligned}
&(\Delta t_k)^{-1}(e_k - e_{k-1})(-\nabla^2 e_k) + \mu_n(\nabla^2 e_k)^2 \\
&\quad = -\mu_n' \nabla^2 u_k \nabla e_k \nabla^2 e_k + \mu_n \nabla u_k \nabla e_k \nabla^2 e_k \\
&\quad + \nabla n \nabla^2 u_k [\mu_n'(\nabla u_k) - \mu_n'(\nabla u)] \nabla^2 e_k \\
&\quad - \nabla n [\nabla^2 u - \nabla^2 u_k] \mu_n'(\nabla u) \nabla^2 e_k - \nabla n [\mu_n(\nabla u_k) \nabla u_k \\
&\quad - \mu_n(\nabla u) \nabla u] \nabla^2 e_k + [\mu_n(\nabla u_k) - \mu_n(\nabla u)] \nabla^2 n \nabla^2 e_k \\
&\quad - V_{n,k} \nabla^2 e_k + r_{2,k} \nabla^2 e_k - f_k \nabla^2 e_k .
\end{aligned}
\tag{165}
$$

Here, f_k and $V_{n,k}$ have the meaning of (86) and (83), respectively. Integrate (165) over Ω, integrating the first term by parts, and utilize (91) with $\lambda = 9/\inf \mu_n$. We sum over $k = 1, \ldots, L$ to obtain

$$
\begin{aligned}
&(\Delta t_L)^{-1} \|\nabla e_L\|_{L^2}^2 + \frac{1}{2} \inf \mu_n \sum_{k=1}^{L} \|\nabla^2 e_k\|_{L^2}^2 \\
&\quad \le \frac{9}{2 \inf \mu_n} \sum_{k=1}^{L} [(v_{sn}^2 + \mu_{0n}^4 \sup |\nabla^2 u_k|^2 / v_{sn}^2) \|\nabla e_k\|_{L^2}^2 \\
&\quad + C(\|e_k\|_{L^2}^2 + \|q_k\|_{L^2}^2) + \|r_{2,k}\|_{L^2}^2 + \|f_k\|_{L^2}^2 \\
&\quad + B^2(\|e_k\|_{L^2}^2 + \|q_k\|_{L^2}^2 + \|r_{1,k}\|_{L^2}^2)]
\end{aligned}
\tag{166}
$$

where

(167)

$$
\begin{aligned}
B = {} & \frac{2(b-a)\mu_{0n}}{\pi\sqrt{\varepsilon}} \left(\frac{\mu_{0n}^2}{v_{sn}^2} \sup |\nabla n| \sup |\nabla^2 u_k| + 2 \sup |\nabla n| + \frac{\mu_{0n}}{v_{sn}} \sup |\nabla^2 n| \right) \\
& + \left(\frac{2\mu_{0n}^2}{\varepsilon v_{sn}} \right) \sup |\nabla n|,
\end{aligned}
$$

and C is a constant arising from the fact that $V_{n,k}$ can be written as a sum of terms exhibiting linear growth in e_k and q_k (cf. (75) and (76)). Note that we have used the Lipschitz properties of $\xi \mapsto \mu_n(\xi)$, $\xi \mapsto \mu_n'(\xi)$, and $\xi \mapsto \xi \mu_n'(\xi)$. Use of the auxiliary estimate,

$$
\|\nabla^2 u_k\|_{L^\infty} \le \varepsilon^{-1}(\|n_k\|_{L^\infty} + \|p_k\|_{L^\infty} + \|N\|_{L^\infty} + \|r_{1,k}\|_{L^\infty}),
\tag{168}
$$

and the hypotheses of the Lemma, now yield the inequality

$$\begin{aligned}(169)\quad &\sum_k (\|\nabla^2 e_k\|_{L^2}^2 + \|\nabla^2 q_k\|_{L^2}^2) \\ &\qquad \le C_1 \sum_k \|r_k\|_{L^2}^2 + \|\mathscr{P}\|(\|n_{tt}\|_{L^2(\Omega\times(0,T_0))}^2 + \|p_{tt}\|_{L^2(\Omega\times(0,T_0))}^2).\end{aligned}$$

The conclusion (159) now follows from (168) and (169) and the specification of the boundary conditions. □

Our next result details the estimation of the Lipschitz constant for F'.

PROPOSITION 8. *For the domain of definition of F restricted to a ball of radius r_0 in X_s, a constant M_3 exists, depending only on r_0, in the case $s = 2$, such that* (135) *holds. In the case $s = 1$, the hypotheses of Lemma 7 are also required, as well as a generalized inverse inequality*:

$$(170)\qquad \|\nabla n_1 - \nabla n_2\|_{L^\infty} \le C_{\|\mathscr{P}\|}\|\nabla n_1 - \nabla n_2\|_{L^2},$$

$$(171)\qquad \|\nabla p_1 - \nabla p_2\|_{L^\infty} \le C_{\|\mathscr{P}\|}\|\nabla p_1 - \nabla p_2\|_{L^2}.$$

In this case, M_3 depends upon $\|\mathscr{P}\|$. Inequalities (170) *and* (171) *need only hold in the sense of Remark 2. In the case $s = 1$, hypothesis* (143) *may be substituted for the hypotheses of Lemma 7.*

PROOF. We consider first the case $s = 2$. Thus, we must estimate M_3 in

$$(172)\quad \|[F'(v_1) - F'(v_2)][\phi, \psi, \omega]\|_{L^2} \le M_3\|v_1 - v_2\|_{H^2}\|[\phi, \psi, \omega]\|_{H^2},$$

where

$$(173)\qquad v_1 = [u_1, n_1, p_1], \qquad v_2 = [u_2, n_2, p_2].$$

The first component of $F'(v_1) - F'(v_2)$ is zero, as is evident from (119)–(123). We estimate separately the terms in the second component of the element on the left-hand side of (172), and we begin by estimating

$$\begin{aligned}&\nabla\cdot[\nabla n_1 \mu_n'(\nabla u_1)\nabla\phi] - \nabla\cdot[\nabla n_2 \mu_n'(\nabla u_2)\nabla\phi] \\ &\quad = \nabla^2(n_1 - n_2)\mu_n'(\nabla u_1)\nabla\phi + \nabla^2 n_2[\mu_n'(\nabla u_1) - \mu_n'(\nabla u_2)]\nabla\phi \\ (174)\qquad &\quad + \nabla(n_1 - n_2)\mu_n''(\nabla u_1)\nabla^2 u_1 \nabla\phi \\ &\quad + \nabla n_2[\mu_n''(\nabla u_1)\nabla^2 u_1 - \mu_n''(\nabla u_2)\nabla^2 u_2]\nabla\phi \\ &\quad + \nabla(n_1 - n_2)\mu_n'(\nabla u_1)\nabla^2\phi + \nabla n_2[\mu_n'(\nabla u_1) - \mu_n'(\nabla u_2)]\nabla^2\phi.\end{aligned}$$

We find the norm of this expression is bounded by

(175)

$$\left(\frac{\mu_{0n}^2}{v_{sn}}\right)\Big\{\|n_1 - n_2\|_{H^2}\|\nabla\phi\|_{L^2} + \left(\frac{\mu_{0n}}{v_{sn}}\right)\|\nabla\phi\|_{L^\infty}[\|n_2\|_{H^2}\|\nabla(u_1 - u_2)\|_{L^2}$$
$$+ \|u_1\|_{H^2}\|\nabla(n_1 - n_2)\|_{L^2} + \|\nabla n_2\|_{L^2}\|u_1 - u_2\|_{H^2}]$$
$$+ \left(\frac{15\mu_{0n}^3}{v_{sn}^2}\right)\|\nabla n_2\|_{L^\infty}\|\nabla^2 u_2\|_{L^2}\|\nabla(u_1 - u_2)\|_{L^2}\|\nabla\phi\|_{L^\infty}$$
$$+ \|\nabla(n_1 - n_2)\|_{L^2}\|\phi\|_{H^2} + \left(\frac{\mu_{0n}}{v_{sn}}\right)\|\nabla n_2\|_{L^2}\|\nabla(u_1 - u_2)\|_{L^\infty}\|\phi\|_{H^2}\Big\}.$$

These terms are clearly of the form given by the right-hand-side estimate of (172). The constant M_3 is seen to depend on r_0. The next two terms to estimate are given by

$$\nabla n_1 \nabla u_1 \mu_n'(\nabla u_1)\nabla\phi - \nabla n_2 \nabla u_2 \mu_n'(\nabla u_2)\nabla\phi$$

and

$$\nabla n_1 \mu_n(\nabla u_1)\nabla\phi - \nabla n_2 \mu_n(\nabla u_2)\nabla\phi.$$

Each of these two differences can be represented analogously to (174). The first difference replaces ∇n_1 and ∇n_2 by n_1 and n_2, respectively, and $\xi \mapsto \mu_n'(\xi)$ by $\xi \mapsto \xi\mu_n'(\xi)$, in the first two terms of the right-hand side of (174). The latter replacement simply means that, in (175), a bound for μ_n' is replaced by a bound for $\xi\mu_n'$; this bound is simply μ_{0n}. The second difference is handled even more easily. The estimation of the (2, 1) block differences is completed by consideration of the terms

$$\begin{aligned}\varepsilon^{-1}\{&2[\mu_n'(\nabla u_1) - \mu_n'(\nabla u_2)]\\ &+ [\mu_n''(\nabla u_1)\nabla u_1 - \mu_n''(\nabla u_2)\nabla u_2]\}(n_1^2 - n_1 p_1 - n_1 N)\\ &+ \varepsilon^{-1}[2\mu_n'(\nabla u_2) + \mu_n''(\nabla u_2)\nabla u_2][(n_1^2 - n_2^2) - (n_1 p_1 - n_2 p_2)\\ &\qquad - (n_1 - n_2)N].\end{aligned}$$

Estimates for these terms in the matrix product are provided by

$$\left(\frac{5\mu_{0n}^3}{\varepsilon v_{sn}^2}\right)\|\nabla(u_1 - u_2)\|_{L^2}\|n_1^2 - n_1 p_1 - n_1 N\|_{L^\infty}\|\nabla\phi\|_{L^2}$$
$$+ \left(\frac{3\mu_{0n}^2}{v_{sn}}\right)\|\nabla\phi\|_{L^2}(\|n_1 - n_2\|_{L^2}\|n_1 + n_2\|_{L^\infty} + \|n_1 - n_2\|_{L^2}\|p_1\|_{L^\infty}$$
$$+ \|p_1 - p_2\|_{L^2}\|n_2\|_{L^\infty} + \|n_1 - n_2\|_{L^2}\|N\|_{L^\infty}).$$

Again this fits into the format of (172).

From the (2, 2) block, we estimate the difference

$$\begin{aligned}&\nabla[\mu_n(\nabla u_1)\nabla\psi] - \nabla[\mu_n(\nabla u_2)\nabla\psi]\\ &\quad = [\mu_n(\nabla u_1) - \mu_n(\nabla u_2)]\nabla^2\psi + \mu_n'(\nabla u_1)\nabla^2 u_1\nabla\psi - \mu_n'(\nabla u_2)\nabla^2 u_2\nabla\psi\end{aligned}$$

as

$$\begin{aligned}\left(\frac{\mu_{0n}^2}{v_{sn}}\right)\Bigg[&\|\nabla(u_1 - u_2)\|_{L^2}\|\psi\|_{H^2}\\ &+ \left(\frac{\mu_{0n}^2}{v_{sn}}\right)\|\nabla(u_1 - u_2)\|_{L^2}\|\nabla^2 u_1\|_{H^2}\|\nabla\psi\|_{L^2}\\ &+ \|u_1 - u_2\|_{H^2}\|\nabla\psi\|_{L^2}\Bigg],\end{aligned}$$

which is again of the form (172). The difference

$$\mu_n(\nabla u_1)\nabla u_1\nabla\psi - \mu_n(\nabla u_2)\nabla u_2\nabla\psi$$

is estimated by

$$2\mu_{0n}\|\nabla(u_1 - u_2)\|_{L^2}\|\nabla\psi\|_{L^2},$$

which is of the form (172). Thc terms

$$\left[\frac{\partial U_n}{\partial n}(u_1, n_1, p_1) - \frac{\partial U_n}{\partial n}(u_2, n_2, p_2)\right]\psi$$

and

$$\left[\frac{\partial U_p}{\partial p}(u_1, n_1, p_1) - \frac{\partial U_p}{\partial p}(u_2, n_2, p_2)\right]\omega$$

are estimated by techniques similar to the above. This concludes the proof for $s = 2$.

In the case $s = 1$, the estimation of the linear functional,

$$(F'[u_1, n_1, p_1] - F'[u_2, n_2, p_2])[\phi, \psi, \omega],$$

proceeds by estimating terms from the (2, 1) block differences beginning with

$$\int_\Omega (\nabla n_1 - \nabla n_2)\mu_n'(\nabla u_1)\nabla\phi\nabla\psi_0 + \int_\Omega \nabla n_2(\mu_n'(\nabla u_1) - \mu_n'(\nabla u_2))\nabla\phi\nabla\psi_0,$$

where the functional acts on the test function $[\phi_0, \psi_0, \omega_0]$. This can be estimated by

$$\begin{aligned}\left(\frac{\mu_{0n}^2}{v_{sn}}\right)&\|\nabla(n_1 - n_2)\|_{L^\infty}\|\nabla\phi\|_{L^2}\|\nabla\psi_0\|_{L^2}\\ &+ \left(\frac{\mu_{0n}^3}{v_{sn}^2}\right)\|\nabla n_2\|_{L^\infty}\|\nabla(u_1 - u_2)\|_{L^\infty}\|\nabla\phi\|_{L^2}\|\nabla\psi_0\|_{L^2}.\end{aligned}$$

This is the most delicate of the estimates, since it involves $\|\nabla(n_1 - n_2)\|_{L^\infty}$ and $\|\nabla n_2\|_{L^\infty}$. The latter can be estimated by Lemma 7, or alternatively (143). The former requires an estimate of the form

$$\|\nabla(n_1 - n_2)\|_{L^\infty} \leq C_{\Delta t}\|\nabla(n_1 - n_2)\|_{L^2},$$

which is a property of the inverse hypothesis (170). The next two terms in the (2, 1) block difference do not require special hypotheses, while the final term in the (2, 1) block may make use of the estimate for $s = 2$. The terms in the (2, 2) and (2, 3) blocks are routinely estimated. □

4.4. *Interface with the inner iteration.* In Part II of this series, we shall examine decoupling inner iterations, giving rise to approximate inverses of the derivative maps $F'(u, n, p)$. Detailed analysis of typical decouplings of this type suggests that the approximate inverse satisfies an inequality of the form

$$\|[(F' \circ G)(u_k^{m-1}, n_k^{m-1}, p_k^{m-1}) - I]r^{m-1}\|_{H^{s-2}} \leq C\Delta t_k \|r^{m-1}\|_{H^s}. \tag{176}$$

Once again, r^{m-1} denotes the residual, at the kth time step, of the $(m-1)$th Newton iteration. The conjunction of (176) and (142) suggests the inequality

$$\|G(u_k^{m-1}, n_k^{m-1}, p_k^{m-1})r^{m-1}\|_{H^s} \leq M(1 + C\Delta t_k)\|r^{m-1}\|_{H^{s-2}}, \tag{177}$$

leading to the choice of M_1 in (133), dictated by

$$M_1 \stackrel{\text{def}}{=} M(1 + C\|\mathscr{P}\|). \tag{178}$$

Here M is defined in (142). To draw the circle completely around (133)–(135), we can identify M_2 of (134) with M_1 and M_3 by

$$M_2 \stackrel{\text{def}}{=} M_1^2 M_3, \tag{179}$$

and then link this choice of M_2 with that induced by (176):

$$M_1^2 M_3 \|r^{m-1}\|_{H^{s-2}} \geq C\Delta t_k. \tag{180}$$

The choice of M_2 in (179) is motivated by (140). Clearly, (180) is implied by

$$M^2 M_3 \|r^{m-1}\|_{H^{s-2}} \geq C\|\mathscr{P}\|, \tag{181}$$

for each $m \leq m_0$. Inequality (181) suggests that a small number of Newton iterations is desirable, to minimize the restriction on $\|\mathscr{P}\|$, provided this is consistent with the residual tolerance described by Theorems 3 and 4. If, as motivated by those theorems, we desire

$$\|r^m\|_{H^s} \leq c\Delta t_k, \tag{182}$$

for specified c, then (138) shows that

$$c\Delta t_k \geq (2M_1^2 M_3)^m \|r^0\|_{H^{s-2}}^{2m} \tag{183}$$

must be satisfied.

Condition (183) imposes an implicit condition on $\|r^0\|_{H^{s-2}}$. For example, if only one Newton iteration is desired, then

$$\|r^0\|_{H^{s-2}} \leq \sqrt{\frac{c\Delta t_k}{2M_1^2 M_3}}. \tag{184}$$

One is tempted to use as a starting guess in the Newton iteration the accepted approximation at the previous time step. Such a selection may fail to satisfy conditions such as (183) or (184). In the next section we briefly discuss suggested procedures, specifically continuation, to achieve this.

4.5. *Bridging time steps.* In an actual implementation, one might select a second-order extrapolation method, based upon the accepted approximations at the two preceding time steps, as a way of determining the starting iterate. Since this depends on the fully discrete algorithm which we shall consider in the sequel to this paper (Part II), we prefer to describe a continuation procedure, which is amenable to the modularity inherent in our algorithmic analysis.

For simplicity, denote the map (118) at the $(k-1)$th time step by F_{k-1} and the map at the kth time step by F_k. By the criterion established in Theorem 3 or Theorem 4, if an approximation triple z_{k-1} is accepted at the $(k-1)$th time step, then $F_{k-1}(z_{k-1})$ is of magnitude $c\Delta t_{k-1}$, in an appropriate norm. We shall allow F_{k-1} and F_k to act, more generally, on elements of the form

$$v_\lambda = v + (1-\lambda)\overline{z}_{k-1} + \lambda \overline{z}_k,$$

where $v \in Y_s$ (cf. (131)) and $\overline{z}_{k-1}$, $\overline{z}_k$ assume the boundary values. Define the homotopy map,

$$F(v,\lambda) = (1-\lambda)F_{k-1}(v_\lambda) + \lambda F_k(v_\lambda), \qquad v \in Y_s,\ 0 \leq \lambda \leq 1. \tag{185}$$

Knowing that $F_{k-1}(z_{k-1})$ is of magnitude $c\Delta t_{k-1}$, we wish to vary λ, introducing intermediate approximations thereby, so that at termination, z_k^0 is obtained, where $F_k(z_k^0)$ satisfies, say, (184), or the more general residual condition (183). Here z_k^0 is a starting guess for a Newton outer iteration. A strategy consistent with the framework of the paper is a predictor/delayed-corrector method. Roughly, assuming

$\Delta t_{k-1} = \Delta t_k = \Delta t$, we have $\lfloor N/2 \rfloor$ for the number of Euler predictor iterations performed at successive λ-values λ_i, before a series of m Newton corrector iterations is performed at some λ_i. Here

$$N^{-1} = (c\Delta t)^{1-1/2^m}(2M_1^2 M_3)^{m/2^m}. \tag{186}$$

The effect of the corrector iterations is to restore the residual to the order $c\Delta t$. We shall present the theory underlying these statements.

The map defined by (185) satisfies the algebraic relation

$$F(v,\lambda) = F(w,\mu) + F'_w(w,\mu)(v-w) + F'_\mu(w,\mu)(\lambda-\mu) + R(v,w;\lambda,\mu), \tag{187}$$

where

$$\begin{aligned} R(v,w;\lambda,\mu) \stackrel{\text{def}}{=} &(1-\mu)\{F_{k-1}(v_\lambda) - F_{k-1}(w_\mu) - F'_{k-1}(w_\mu)(v_\lambda - w_\mu)\} \\ &+ \mu\{F_k(v_\lambda) - F_k(w_\mu) - F'_k(w_\mu)(v_\lambda - w_\mu)\} \\ &+ (\mu-\lambda)\{[F_{k-1}(v_\lambda) - F_{k-1}(w_\mu)] - [F_k(v_\lambda) - F_k(w_\mu)]\}. \end{aligned} \tag{188}$$

If hypothesis (135) is satisfied, then (187) and (188) imply the smoothness estimate

$$\|R(v,w;\lambda,\mu)\| \le c_1\|v_\lambda - w_\mu\|^2 + c_2(\lambda-\mu)^2. \tag{189}$$

The hypotheses which are analogous to, but weaker than, (133) and (134) are:

$$\|H(v,\lambda)z\| \le M'_1\|z\|, \tag{190}$$

$$\|[F'_v(v,\lambda)H(v,\lambda) - I]z\| \le M'_2\|z\|, \tag{191}$$

for $z = F'_\lambda(v,\lambda)$, where $\{H(v,\lambda)\}$ comprises a set of approximate right inverses for $F'_v(v,\lambda)$. Note that $\|z\|$, not $\|z\|^2$, appears in (191); also, notice that $F'_\lambda(v,\lambda)$ has the representation

$$F'_\lambda(v,\lambda) = [F_k(v_\lambda) - F_{k-1}(v_\lambda)] + [\lambda F'_k(v_\lambda) + (1-\lambda)F'_{k-1}(v_\lambda)](\overline{z}_k - \overline{z}_{k-1}). \tag{192}$$

The control of $|\Delta\lambda|\|F'_\lambda(v,\lambda)\|$ is critical for the continuation, as v_{i-1} is replaced by a new predictor v_i. In practice, $\Delta\lambda$ is chosen adaptively, but here we assume for simplicity that the λ-points are equally spaced. We have the following.

PROPOSITION 9. *Let*

$$\rho = \|F_{k-1}(z_{k-1})\| \le c\Delta t_{k-1}, \tag{193}$$

suppose that $z_{k-1} \in B_{\alpha r_0}$, $0 \le \alpha < 1$, *and let* $\Delta\lambda$ *be specified, with* $\lambda_i \stackrel{\text{def}}{=} i\Delta\lambda$. *Here* B_{r_0} *is a ball in* H^2 *of radius* r_0. *Suppose, for simplicity, that* $\Delta t_{k-1} = \Delta t_k = \Delta t$ *and that the successive Euler predictors are defined by* $v_0 \stackrel{\text{def}}{=} z_{k-1}$,

$$\frac{v_i - v_{i-1}}{\Delta\lambda} = -H(\hat{v}_{i-1}, \lambda_{i-1}) F'_\lambda(\hat{v}_{i-1}, \lambda_{i-1}), \quad \textit{for each } i. \tag{194}$$

Here, $\hat{v}_i$ *denotes the affine translate of* v_i *such that* $\hat{v}_i \in Y_s$. *Then, if* $\Delta\lambda$ *is such that* (197) *does not exceed* $c\Delta t$, *and such that*

$$|\Delta\lambda| M'_1 \|F'_\lambda(\hat{v}_{i-1}, \lambda_{i-1})\| \leq 2(1-\alpha) r_0 / N, \tag{195}$$

where N *is given by* (186), *then it follows that* $v_1, \ldots, v_{\lfloor N/2 \rfloor}$ *are in* B_{r_0}, *and* $v_{\lfloor N/2 \rfloor}$ *satisfies the residual condition* (183). *In particular, the mth Newton iterate, starting with* $v_{\lfloor N/2 \rfloor}$ *as the zeroth iterate, again has residual of magnitude* $c\Delta t$.

PROOF. By direct estimation, if w_0 denotes the center of B_r, we have, for $i = 1, \ldots, \lfloor N/2 \rfloor$,

$$\begin{aligned} \|v_i - v_0\| &\leq \|v_i - v_{i-1}\| + \|v_{i-1} - w_0\| + \|v_0 - w_0\| \\ &\leq |\Delta\lambda| \|H(\hat{v}_{i-1}, \lambda_{i-1})\| \|F'_\lambda(\hat{v}_{i-1}, \lambda_{i-1})\| \\ &\qquad\qquad + 2(i-1)(1-\alpha) r_0 / N + \alpha r_0 \\ &\leq 2i(1-\alpha) r_0 / N + \alpha r_0 \end{aligned}$$

if (195) and the induction hypotheses are invoked. It follows that $v_1, \ldots, v_{\lfloor N/2 \rfloor}$ are in B_{r_0}.

For the residual we have

$$\begin{aligned} F(\hat{v}_i, \lambda_i) = & -\Delta\lambda [F'_v(\hat{v}_{i-1}, \lambda_{i-1}) H(\hat{v}_{i-1}, \lambda_{i-1}) - I] \circ F'_\lambda(\hat{v}_{i-1}, \lambda_{i-1}) \\ & + R(\hat{v}_i, \hat{v}_{i-1}; \lambda_i, \lambda_{i-1}) + F(\hat{v}_{i-1}, \lambda_{i-1}). \end{aligned} \tag{196}$$

The estimation of the first two terms is given by

$$|\Delta\lambda| M'_2 \|F'_\lambda(\hat{v}_{i-1}, \lambda_{i-1})\| \{1 + c_1 M'_2 |\Delta\lambda| \|F'_\lambda(\hat{v}_{i-1}, \lambda_{i-1})\|\} + c_2 |\Delta\lambda|^2. \tag{197}$$

By the hypothesis on $\Delta\lambda$, this quantity does not exceed $c\Delta t$. It follows from this analysis that the residual corresponding to $v_{\lfloor N/2 \rfloor}$ does not exceed $Nc\Delta t$. By direct computation, this satisfies (183). □

5. Postscript. This paper has described a modular algorithm development beginning with the semidiscrete systems determined by a fully implicit time discretization. Though first-order methods are analyzed here, the ideas are capable of extension to higher-order time discretizations.

The modular development proceeds from the fundamental observation, contained in Theorems 3 and 4, that the accuracy of Euler's method is maintained by approximate solvability, as specified by the residuals. The approximate solutions will not, in general, satisfy the invariant-region principle developed for the exact semidiscrete solution in Theorem 1, though this theory provides the appropriate analytical infrastructure, as well as a guide to the properties desired of a computational procedure.

Newton's method is our choice for approximate solvability. If the approximate solvability condition is expressed in terms of a residual, numerically bounded in norm by $c\Delta t_k$, and if m is the number of outer Newton iterations at t_k, then inequality (183) determines the necessary condition on the starting guess at the commencement of the outer iterations at $t = t_k$. The constants M_1 and M_3 in this inequality are associated with a class of generic approximate Newton methods. Their relation to the semiconductor model is drawn via (178) and Propositions 6 and 8. Although the constants appear not to be completely explicit, the existential components can be eliminated by a careful correlation between the radius of the local domain of definition of F and gradient estimates induced by Sobolev-type inequalities in one dimension.

It is natural to wish to use the accepted approximation at the previous time step as the starting guess at the current time step. Such a starting iterate, however, may fail to satisfy (183), if m is maintained as a small positive integer, e.g., $m = 1$ or $m = 2$; the potential boundary conditions are clearly not satisfied either. An approximate time-stepping bridge is furnished by a continuation method, specifically, a predictor-corrector method as described in Proposition 9. The analytical framework permits several Euler predictor steps, before approximate Newton steps are applied.

The function-space framework, in which the approximate Newton methods are analyzed at the differential equation level, admits both a classical and variational formulation. These are loosely organized via L^2 and H^{-1} settings. The latter requires a significant amount of technical supporting analysis; e.g., the $s = 1$ subcase of Proposition 6 requires the companion Lemma 7. Both the settings have been included with an eye to the fully discrete algorithms to be developed in the sequel to this paper, Part II.

We have deliberately refrained from specifying the specific form of the approximate Newton method. In fact, we view the exact Newton iterations as *outer iterations*, each realized in terms of a number of so-called inner iterations. The sum total of these inner iterations, corresponding to each outer iteration, represents the approximate Newton iteration; this process is repeated m times at each time step. The inner iterations are visualized as cumulative, in the sense that each succeeding iteration enhances the previous; also, it must be fully discrete in its final form. Our numerical studies thus far have indicated that the substitution giving rise to (14) and (15) creates a "de facto" scale imbalance in the Jacobian; in order for a stable computation to proceed, some time-step restriction must be imposed. We shall deal with these issues in Part II, where a specific class of algorithms together with numerical studies, will be presented.

References

1. R. E. Bank, W. M. Coughran, Jr., W. Fichtner, E. H. Grosse, D. J. Rose, and R. K. Smith, *Transient simulation of silicon devices and circuits*, IEEE Trans. CAD- **4** (1985), 436–451.

2. R. E. Bank, J. W. Jerome, and D. J. Rose, *Analytical and numerical aspects of semiconductor device modeling*, Computing Methods in Applied Sciences and Engineering, V (1982), 593–597 (R. Glowinski and J.-L. Lions, eds.), North-Holland.

3. R. E. Bank and D. J. Rose, *Global approximate Newton methods*, Numer. Math. **37** (1981), 279–295.

4. R. E. Bank, D. J. Rose, and W. Fichtner, *Numerical methods for semiconductor device simulation*, IEEE Trans. ED **30** (1983), 1031–1041.

5. A. Berman and R. J. Plemmons, *Nonnegative Matrices in the Mathematical Sciences*, Academic Press, 1979.

6. D. M. Caughey and R. E. Thomas, *Carrier mobilities in silicon empirically related to doping and field*, Proc. IEE-E **55** (1967), 2192–2193.

7. A. DeMari, *An accurate numerical one-dimensional solution of the P-N junction under arbitrary transient conditions*, Solid-State Electron **11** (1968), 1021–1053.

8. A. DeMari, *An accurate numerical steady-state one-dimensional solution of the P-N junction*, Solid-State Electron **11** (1968), 33–58.

9. D. Gilbarg and N. S. Trudinger, *Elliptic Partial Differential Equations of Second Order*, Springer-Verlag, 2nd edition, 1983.

10. R. N. Hall, *Electron-hole recombination in germanium*, Phys. Rev. **87** (1952), 387.

11. J. W. Jerome, *Approximate Newton methods and homotopy for stationary operator equations*, Constr. Approx. **1** (1985), 271–285.

12. J. W. Jerome, *Approximation of Nonlinear Evolution Systems*, Academic Press, 1983.

13. J. W. Jerome, *Consistency of semiconductor modeling: an existence/stability analysis for the stationary van Roosbroeck system*, SIAM J. Appl. Math. **45** (1985), 565–590.

14. J. W. Jerome, *Evolution systems in semiconductor device modeling: a cyclic uncoupled line analysis for the Gummel map*, Math. Methods Appl. Sci. **9** (1987), 455–492.

15. T. Kato, *The Cauchy problem for quasi-linear symmetric hyperbolic systems*, Arch. Rat. Mech. Anal **58** (1975), 181–205.

16. P. A. Markowich, *The Stationary Semiconductor Device Equations*, Springer-Verlag, 1986.

17. R. P. Mertens, R. J. van Overstraeten, and H. J. de Man., *Heavy doping effects in silicon*, Advances in Electronics and Electron Phys. **55** (1981), 77–118.

18. W. Shockley and W. T. Read, *Statistics of the recombination of holes and electrons*, Phys. Rev. **87** (1952), 835–842.

19. S. M. Sze, *Physics of Semiconductor Devices*, Wiley-Interscience, 2nd edition, 1981.

20. K. K. Thornber, *Relation of drift velocity to low-field mobility and high-field saturation velocity*, J. Appl. Phys. **51** (1980), 2127–2136.

21. W. Van Roosbroeck, *Theory of flow of electrons and holes in germanium and other semiconductors*, Bell System Tech. J **29** (1950), 560–607.

22. R. S. Varga, *Matrix Iterative Analysis*, Prentice-Hall, 1962.

COMPUTING MATHEMATICS RESEARCH DEPARTMENT, AT&T BELL LABORATORIES, MURRAY HILL, NEW JERSEY 07974

NORTHWESTERN UNIVERSITY, EVANSTON, ILLINOIS 60201

Lectures in Applied Mathematics
Volume **25** (1990)

Efficiency and Acceleration of Steady-State Decoupling Algorithms

THOMAS KERKHOVEN

1. Introduction. The nonlinear block Gauß-Seidel algorithm, known in semiconductor simulation as "Gummel's method," has a long-standing popularity for the solution of the equations describing steady state semiconductors. Originally it was proposed by H. K. Gummel in 1964 **[6]** for solution of the one-dimensional steady-state equations. In 1969 J. Slotboom **[15]** generalized this approach to two-dimensional problems.

Instead of expressing the semiconductor in terms of the physical variables u, n, and p we transform the densities n and p to the quasi-Fermi levels v and w defined through $e^{u-v} = n$ and $e^{w-u} = p$. We assume constant mobilities, Einstein's relations and zero generation and recombination terms in the current continuity equations. In terms of the electrostatic potential u and the quasi-Fermi levels v and w a steady-state semiconductor is then described by the system of nonlinear boundary value problems below:

$$
\begin{aligned}
-\nabla^2 u + e^{u-v} - e^{w-u} - k_1 &= 0, \\
-\nabla \cdot (e^u \nabla e^{-v}) &= 0, \\
-\nabla \cdot (e^{-u} \nabla e^{w}) &= 0.
\end{aligned}
\tag{1}
$$

Analogously to **[10]**, **[13]**, **[11]**, we define an iterative decoupling algorithm for the solution of the system above through the successive application of a fixed-point mapping $T_F : u \to \tilde{u}$ in which first either

1980 *Mathematics Subject Classification* (1985 *Revision*). Primary 65C20, 35J60, 47H12, 35J65.

one or both current continuity equations are solved for the intermediate variable v or both v and w and subsequently the potential equation is solved for the image $\tilde{u}$. Thus T_F is defined by the set of boundary value problems

$$\begin{aligned} -\nabla \cdot (e^{u-v}\nabla v) &= 0, \\ -\nabla \cdot (e^{-u+w}\nabla w) &= 0, \\ -\nabla^2 \tilde{u} + e^{\tilde{u}-v} - e^{w-\tilde{u}} - k_1 &= 0. \end{aligned} \tag{2}$$

Existence of a fixed point of the mapping T_F, which corresponds to a solution of the equations (1), follows from the existence proof for an analogous mapping from pairs (v, w) to their images $(\tilde{v}, \tilde{w})$ which are chosen to be intermediate variables now, presented by Jerome in [**8**].

2. The effectiveness of the decoupling approach. Even if Gummel's method does not converge sufficiently fast in many cases, it is found to be remarkably successful for its conceptual simplicity. Especially for small variation of the "bias potentials," i.e., small variation in the boundary conditions for the quasi-Fermi levels v and w, the method converges quickly. For small variation of the boundary data for v and w this may not be all that surprising because v and w satisfy maximum principles by which they cannot exceed the extrema which they assume at the Dirichlet part of the boundary. Hence the system decouples completely for zero variation of these boundary data because in that case v and w are constants irrespective of u. However, even for large variation of the boundary data for v and w Gummel's method is still found to converge, albeit more slowly. Moreover, Gummel's method converges more quickly while the iterates are far away from the solution than when they are close. In the sequel we will first discuss the global behaviour of the algorithm, far away from the solution, and subsequently the local behaviour close to the solution.

3. Contraction-mapping analysis. In this section we analyze the convergence of Gummel's method by contraction-mapping analysis. Thus we find conditions under which the fixed-point mapping T_F is a contraction as defined in the following definition.

DEFINITION 1. A mapping $T: X \to X$, where X is a subset of a normed linear space N, is called a *contraction mapping*, or simply a *contraction*, if there is a positive number $C < 1$ such that

$$\|Tx - Ty\| \leq C\|x - y\| \qquad \forall x, y \in X.$$

If a mapping T is a contraction, the existence of a unique fixed point of T in X follows by the contraction mapping theorem as stated, for instance, in Theorem 5.15 from **[5]**:

THEOREM 1 (CONTRACTION-MAPPING THEOREM). *If $T\colon X \to X$ is a contraction mapping of a closed subset X of a Banach space, then there is exactly one $x^{(*)} \in X$ such that $Tx^{(*)} = x^{(*)}$. For any $x^{(0)} \in X$, the sequence $\{x^{(n)}\}$ defined by $x^{(n+1)} = Tx^{(n)}$ converges to $x^{(*)}$.*

In **[13]** we analyze convergence with respect to the norm L_∞ through contraction-mapping analysis. Convergence with respect to the norm L_∞ follows by demonstrating that the mapping T_F is a contraction with respect to this norm. However, apart from convergence for sufficiently small variation of the boundary data for v and w we obtained, moreover, that iterates generated by the decoupling algorithm converge unconditionally while the distance to the solution is sufficiently large. Once the iterates are close to the solution, on the other hand, the analysis no longer guarantees convergence. The situation is summarized in the following theorem which is rephrased from **[13]**.

THEOREM 2. *The one-dimensional version of the nonlinear mapping* (2) *maps two originals u_1 and u_2 at distance*

$$D \equiv \|u_2 - u_1\|_{L_\infty}$$

to images $\tilde{u}_1$ and $\tilde{u}_2$ at distance $\widetilde{D}$ for which

$$\widetilde{D} \leq D + 2\log\left[\frac{1 + e^{D+\frac{1}{2}(v_{\max}-v_{\min})}}{e^{\frac{1}{2}D}[e^D + e^{\frac{1}{2}(v_{\max}-v_{\min})}]}\right].$$

This implies that

- *the contraction constant C of the mapping T_F as defined by* (2) *is bounded from above with respect to the norm L_∞ by the expression*

$$1 - \frac{3 - e^{\frac{1}{2}(v_{\max}-v_{\min})}}{e^{-\frac{1}{4}D}[e^D + e^{\frac{1}{2}(v_{\max}-v_{\min})}]},$$

  so that (2) *is a contraction if $v_{\max} - v_{\min} < \log(9)$ and the decoupling algorithm* (2) *converges in the norm L_∞ to the unique solution $(u^{(*)}, v^{(*)}, w^{(*)})$ of* (1).
- *If $v_{\max} - v_{\min} > \log(9)$ the argument of the logarithm is not uniformly smaller than* 1, *but if*

$$D \geq \cosh^{-1}\left[\frac{1}{2}(e^{\frac{1}{2}(v_{\max}-v_{\min})} - 1)\right]$$

then

$$\tilde{D} < D.$$

Because we can take $u_2 = u^{(*)}$ this implies that sufficiently far away from the solution $u^{(*)}$ we will unconditionally improve our estimate.

The rapid convergence far away from the solution and the decrease in speed of convergence as the solution is approached, as follows from the analysis in this theorem, coincide with results found in numerical experiments. However, this contraction-mapping analysis fails to explain the slow but persisting convergence of the algorithm close to the solution. We turn to spectral analysis of the mapping for more precise results.

4. Spectral analysis. Even the strongest contraction mapping result falls short of explaining why Gummel's algorithm is not found to diverge close to the solution. More precisely, the analysis we mentioned in the previous section showed as well that if the variation in the "bias potentials" is bigger than log(9) then two estimates to the potential u at distance D_u in the norm L_∞ can be mapped by the current continuity equations to two pairs of estimates to the quasi-Fermi levels v and w at a distance $D_{v,w}$ which is bigger than D_u if

$$D_u < \cosh^{-1}\left[\frac{1}{2}(e^{\frac{1}{2}(v_{\max}-v_{\min})} - 1)\right].$$

In fact, it is possible to write down the estimates to u for which this bound is assumed. In other words, in the norm L_∞ we have that if the estimates to u are close enough together their images under the current continuity equations can be further apart than the originals. This demonstrates that at least for the submap corresponding to the current continuity equations one cannot expect to show unconditional contractiveness with respect to the norm L_∞.

In order to improve the convergence result in [**13**] we first observe that although contractivity with respect to the norm L_∞ of the fixed-point mapping T defining the iterative algorithm is a sufficient condition for proving convergence, it is not a necessary condition. A weaker sufficient condition for the local rather than global convergence of a nonlinear map T is stated in Kitchen's theorem [**14**] as stated below.

THEOREM 3. *Let T be a mapping whose domain and range are subsets of a Banach space X. Suppose that*

1. *$x^* \in X$ is a fixed point of T,*
2. *T is differentiable at x^*,*
3. *the spectral radius of the derivative of T at x^* is less than* 1.

Then there exists a neighborhood N of x^* such that

$$\lim_{n\to\infty} T^n(x) = x^*$$

for each $x \in N$.

Thus we only need to examine the spectrum of the derivative of T_F at the solution $u^{(*)}$.

Jerome was the first to examine the eigenvalues of the derivative of the map from v to v for the one-carrier model in [7]. Jerome's results in [8] imply that the mapping T_F, and the submap $U: (v, w) \to u$ defined implicitly through Poisson's equation for the potential, the submap $V: u \to v$ defined implicitly through the current continuity equation for the conduction electrons and the mapping $W: u \to w$ defined implicitly through the equation for the holes are all compact. Therefore, as discussed in [**11**], the local convergence of the fixed-point iteration T_F^n to a fixed point u^* depends not on all spectral values, but only on the size of the eigenvalues of the derivative $T_{F,u}(u^*)$.

For the general N-dimensional problem we have that the size of the eigenvalues of the derivatives U_v, U_w, V_u, and W_u is unconditionally smaller than 1. The result for the current continuity equation for the electrons is an essential improvement over the contraction result in the previous section. It is stated in the following theorem from [**11**] and is valid for the realistic mixed boundary value problem.

THEOREM 4. *For the derivative V_u of the mapping V defined implicitly through the current continuity equation for the electrons we have that*

- *the eigenvalues λ_n are given by*

$$\lambda_n = \frac{1}{1 - ia_n}$$

where $a_n \neq 0$ and real. Thus the λ_n are on the circle in the complex plane which is centered at $1/2$ and has radius $1/2$ while 1 is not an eigenvalue.

- *The eigenfunctions v_n are orthogonal with respect to the inner product*

$$\langle v, w\rangle \equiv \int_0^L e^{u-v} \nabla v \cdot \nabla w^* \, dx.$$

- *The definition of this inner product depends on the point u where the derivative is taken.*
- *The eigenvalues are independent of the device length L.*

For the general N-dimensional case we are not able to prove that for the total mapping T_F we have $\rho[T_{F,u}(x^*)| < 1$ unconditionally. This may imply that there do exist solutions for this system of equations to which Gummel's method will not converge. However, it is not unreasonable to expect local convergence for a larger class of problems from the eigenvalue analysis than from the contraction-mapping analysis because local convergence follows without the uniqueness result as is yielded by the latter. Uniqueness being unlikely on physical grounds, it is an advantage that the spectral convergence analysis finds a way around the uniqueness implication of the contraction-mapping analysis.

The one-dimensional results are more detailed and more complete than the N-dimensional ones. Now the system consists of three coupled two-point boundary value problems for which we assume Dirichlet conditions. For this model the exact location of the eigenvalues of the derivatives of the mappings defined by the current continuity equations can be obtained. For the special one-dimensional situation of zero doping k_1 we can show that the size of all eigenvalues of the derivative $T_{F,u}$ of the mapping T_F is smaller than 1 at the fixed point x^* which yields unconditional local convergence. Furthermore, the speed of convergence of Gummel's method which follows from this special one-dimensional analysis depends on the device length as found before by Jerome in **[9]**.

THEOREM 5. *For the mapping $T: u \to \tilde{u}$ implicitly defined through the system of one-dimensional two-point boundary value problems*

$$\begin{aligned} -[e^{u-v} v_x]_x &= 0, \\ -[e^{w-u} w_x]_x &= 0, \\ -\tilde{u}_{xx} + e^{\tilde{u}-v} + e^{w-\tilde{u}} &= 0, \end{aligned} \tag{3}$$

on a domain $\Omega \equiv [0, L]$ the eigenvalues of the derivative T_u at the solution are determined by the system of homogeneous boundary value problems

$$\begin{aligned} -\mu_{xx} + N(\mu - \nu) + P(\mu - \omega) &= 0, \\ \left[\frac{v(L) - v(0)}{L}\left(\frac{\mu}{\lambda} - \nu\right) + \nu_x\right]_x &= 0, \\ \left[\frac{w(L) - w(0)}{L}\left(\frac{u}{\lambda} - \omega\right) - \omega_x\right]_x &= 0, \end{aligned} \tag{4}$$

and are bounded from above as

$$|\lambda| \leq \frac{N + P}{(\pi/L)^2 + N + P} < 1.$$

Acceleration of the decoupling scheme. The eigenvalue analysis together with experimentally determined spectra provide sufficient information to devise an acceleration scheme for the decoupling algorithm.

The location of the eigenvalues in the complex plane conforms rather accurately to the theoretical results. Thus all eigenvalues are spread throughout the circle in the complex plane with radius $1/2$ centered at $1/2$ and are bounded away from 1. In **[12]** we discuss in detail an acceleration scheme of the decoupling algorithm based on this information. We can increase the convergence rate of the decoupling approach by up to a factor of 5.

The decoupled approach compared to Newton's method. In the previous sections we have discussed the convergence of the nonlinear block Gauß-Seidel approach. The total mapping T_F defining this decoupling algorithm and its submaps do possess some special stability properties which makes the decoupling algorithm more effective than might be expected at first sight. Nevertheless it does not converge quickly for all boundary conditions and may even diverge if certain physical effects are introduced. The modification of Newton's method for the solution of the fully coupled system of PDES as proposed by Bank and Rose [**1**], [**2**], [**3**], [**4**] is a highly successful and more robust approach. Although this is a fast nonlinear algorithm, the solution of the full set of Newton's equations rather than the three subsets at each iteration is relatively expensive. Moreover, Newton's method is more difficult to program. Close to the solution this retrieves quadratic convergence, but far away the modification of Newton's method can be highly ineffective in improving solution estimates.

References

1. R. E. Bank and D. J. Rose, *Analysis of a multilevel iterative method for nonlinear finite element equations*, Math. Comp. **39 (160)** (1982), 453–465.

2. ——, *Global approximate Newton methods*, Numer. Math. **37** (1981), 279–295.

3. ——, *Parameter selection for Newton-like methods applicable to nonlinear partial differential equations*, SIAM J. Numer. Anal. **17 (6)** (1980), 806–822.

4. R. E. Bank, D. J. Rose, and W. Fichtner, *Numerical methods for semiconductor device simulation*, SIAM J. Sci. Statist. Comput. **4 (3)** (1983), 416–435.

5. D. H. Griffel, *Applied Functional Analysis*, Ellis Horwood Limited, 1981.

6. H. K. Gummel, *A self-consistent iterative scheme for one-dimensional steady state transistor calculations*, IEEE Trans. Elec. Dev. **ED-11** (1964), 455–465.

7. J. W. Jerome, *A criterion for isolated solution structure and global computability for operator equations*, J. of Approx. Theory **48 (3)** (1986), 294–302.

8. ——, *Consistency of semiconductor modelling: an existence/stability analysis for the stationary van Roosbroeck system*, SIAM J. Appl. Math. **45 (4)** (1985), 565–590.

9. ——, *The role of semiconductor device diameter and energy-band bending in convergence of Picard iteration for Gummel's map*, IEEE Trans. Elec. Dev. **ED-32(10)** (1985), 2045–2051.

10. T. Kerkhoven, *A proof of convergence of Gummel's algorithm for realistic boundary conditions*, SIAM J. Numer. Anal. **23,** No. 6 (1986), 1121–1137.

11. ——, *A spectral analysis of the decoupling algorithm for semiconductor simulation*, SIAM J. Numer. Anal. **25,** No. 6 (1988), 1299–1312.

12. T. Kerkhoven and Y. Saad, *On acceleration methods for systems of coupled nonlinear partial differential equations*, submitted for publication, February, 1989.

13. T. Kerkhoven, *On the effectiveness of Gummel's method*, SIAM J. Sci. Statist. Comput. **9,** No. 1 (1988), 48–60.

14. J. W. Kitchen, Jr., *Concerning the convergence of iterates to fixed points*, Studia Math. **XXVII** (1966), 247–249.

15. J. W. Slotboom, *Iterative scheme for* 1- *and* 2-*dimensional dc-transistor simulation*, Electron. Lett. **5** (1969), 677–678.

DEPARTMENT OF COMPUTER SCIENCE, UNIVERSITY OF ILLINOIS, URBANA, IL 61801

Lectures in Applied Mathematics
Volume **25** (1990)

Continuation Methods for Parameter-Dependent Boundary Value Problems

HANS D. MITTELMANN

Abstract. An introduction is given into parameter-dependent nonlinear equations exhibiting nonuniqueness of solutions and bifurcation. In particular nonlinear systems of elliptic differential equations are considered. Recently-proposed multigrid continuation methods are surveyed that have been implemented in public domain software. One area of application is the VLSI device simulation. Then related variational inequalities corresponding to obstacle problems are introduced. Theoretical results on continuation are quoted and a method is presented. Spurious transition points are observed in numerical computations.

1. Introduction. Boundary value problems for partial differential equations that describe phenomena in different fields of application are usually nonlinear and depend on one or several parameters of the underlying system. With respect to these parameters nonunicity effects and bifurcation frequently occur. For a list of different problems from the applications as well as an introduction to analytic and numerical aspects see [**18**]. A survey of the state of the art was given in [**15**].

While parameter-dependent problems with known boundaries have been investigated recently by many researchers, only little has been done so far for corresponding problems with a free boundary. Problems that can be written as parameter-dependent variational inequalities are the subject of [7], [**8**], [**9**], [**16**], [**17**], [**21**], [22], [**28**].

Frequently, the only way to explore the solution manifolds of the above problems is to continue along the solution branches with respect

1980 *Mathematics Subject Classification* (1985 *Revision*). Primary 65N25, 65N30.
Supported by the Air Force Office of Scientific Research under Grant AFOSR 84-0315.

to one or very few of the parameters for fixed values of the others. In the case of equations certain natural methods have been proposed and used for continuation (see **[14]**, **[26]**, **[30]**) while for variational inequalities both theoretical as well as numerical results were only obtained recently and these problems are still under investigation.

In the following we will survey some of the continuation methods proposed for both nonlinear equations and systems as well as variational inequalities. To make the presentation more self-contained some of the relevant theoretical results will be developed or quoted, too. For nonlinear systems of partial differential equations the emphasis will be on new continuation methods that have been implemented in public domain software. The field of applications includes, in particular, simulation of VLSI devices.

Many problems in mechanics and other engineering areas but also in other fields as, for example, mathematical economics have in addition to (initial) boundary conditions additional constraints that often correspond to obstacles. The second part of this paper will deal with such problems. Theoretical results about continuation are given as well as an algorithm and some numerical results.

2. Nonlinear eigenvalue problems and bifurcation. In this section a short presentation of some basic facts about the solutions of a general problem

$$G(u,\lambda) = 0, \qquad G\colon X \times \mathbf{R} \to \mathbf{R}, \tag{2.1}$$

will be given in order to provide some background and context for the following sections. X in (2.1) is some function space that will be specified later.

The general problem is to determine all (u, λ) that solve (2.1). Often solutions are sought that satisfy additional equations or conditions or that have certain properties. Of some importance are singular points (u_0, λ_0) or (2.1), i.e.,

$$G(u_0,\lambda_0) = 0, \qquad G_u(u_0,\lambda_0) \text{ singular.} \tag{2.2}$$

The simplest possible case is $X = \mathbf{R}$ and for this we give some very elementary examples.

Fold (*turning*) *points.*

$$G(u,\lambda) = Lu - \lambda e^u = 0, \qquad L \in \mathbf{R}. \tag{2.3}$$

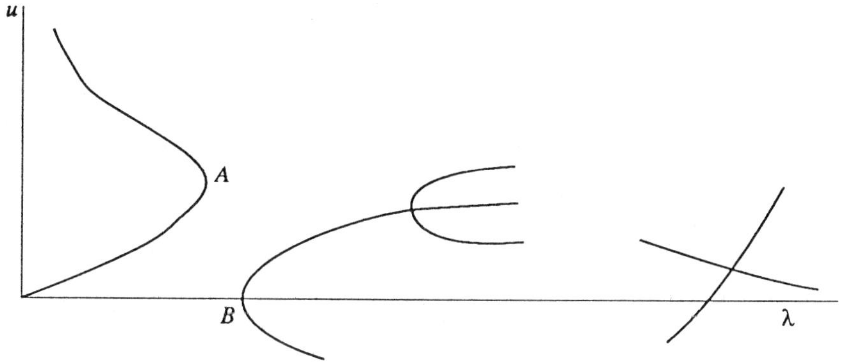

Figure 2.1. Bifurcation diagrams, singular points marked

The singular point (cf. point A in Figure 2.1)

$$(u_0, \lambda_0) = (1, L/e)$$

is easily computed as is the graph of $G^{-1}(0)$. For each λ, $0 < \lambda < \lambda_0$, there are two values of u. This nonuniqueness also implies by the implicit function theorem that G_u must be singular in (u_0, λ_0).

Bifurcation points.

$$G(u, \lambda) = Lu + cu^3 - \lambda u = 0, \qquad L, c \in \mathbf{R},\ c \neq 0. \tag{2.4}$$

There are two solutions

$$u^1(\lambda) = 0, \qquad u^2(\lambda) = \pm\sqrt{(\lambda - L)/c}.$$

The point $(u_0, \lambda_0) = (0, L)$ is a bifurcation point where the solution u^2 "branches" off u^1 (cf. point B in Figure 2.1).

Perturbed bifurcation. In general, physical imperfections are present in the case that (2.1) describes a certain system. If on the other hand such a problem is approximated by, for example, discretization then errors are introduced. Both these phenomena affect the original problem in a similar way. This may be modeled for (2.4) by

$$G(u, \lambda) = Lu + cu^3 - \lambda u - \tau = 0, \qquad \tau \in \mathbf{R}, \tag{2.5}$$

where τ represents either an imperfection or an error term. It is again easy to see that the resulting bifurcation diagram is the one given in Figure 2.2.

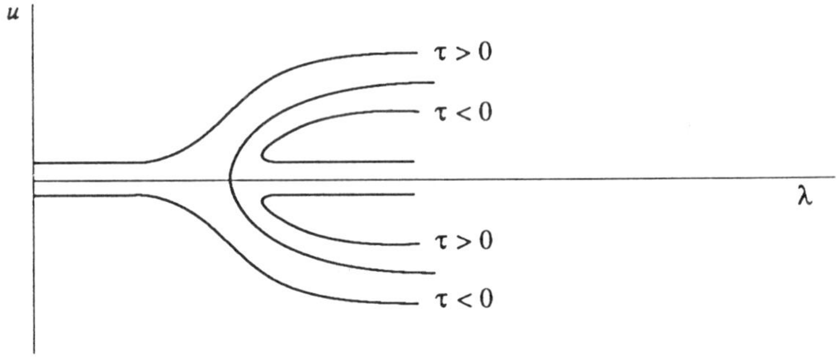

Figure 2.2. Perturbed bifurcation for (2.5)

3. Simple singular points. In order to present a mathematically rigorous framework for at least an important special situation we concentrate on (2.1) with a Banach space X and classify simple singular points (u_0, λ_0).

Let the nullspace of $G_u^0 = G_u(u_0, \lambda_0)$ be spanned by the normalized element φ_1 of X. Let further the range of G_u^0 be closed and given by

$$R(G_u^0) = \{x \in X; \varphi_1^* x = 0\}$$

where φ_1^* spans the nullspace of the adjoint operator. We assume $\varphi_1^* \varphi_1 = 1$.

We now parametrize both u and λ by a parameter s and differentiate (2.1) twice with respect to s (denoted by a dot).

$$\begin{aligned} &G_u^0 \dot{u}_0 + G_\lambda^0 \dot{\lambda}_0 = 0, \\ &G_u^0 \ddot{u}_0 + G_\lambda^0 \ddot{\lambda}_0 = (G_{uu}^0 \dot{u}_0 \dot{u}_0 + 2G_{u\lambda}^0 \dot{u}_0 \dot{\lambda}_0 + G_{\lambda\lambda}^0 \dot{\lambda}_0^2). \end{aligned} \tag{3.1}$$

From the first equation we have

$$\dot{\lambda}_0 d = 0, \qquad d = \varphi_1^* G_\lambda^0.$$

Fold points are characterized by $\dot{\lambda}_0 = 0$, $d \neq 0$, i.e., $G_\lambda^0 \notin R(G_u^0)$. In *bifurcation points* $d = 0$ and $G_\lambda^0 \in R(G_u^0)$. If φ_0 is the solution of

$$G_u^0 \varphi_0 + G_\lambda^0 = 0, \qquad \varphi_1^* \varphi_0 = 0,$$

then

$$\begin{aligned} \dot{u}_0 &= \xi_0 \varphi_0 + \xi_1 \varphi_1, \\ \dot{\lambda}_0 &= \xi_0, \end{aligned}$$

which after inserting into the second equation of (3.1) and multiplying from the left by φ_1^* yields the *bifurcation equation*

$$a\xi_1^2 + 2b\xi_1\xi_0 + c\xi_0^2 = 0 \tag{3.2}$$

where

$$a = \varphi_1^* G_{uu}^0 \varphi_1 \varphi_1,$$
$$b = \varphi_1^* (G_{uu}^0 \varphi_0 + G_{u\lambda}^0) \varphi_1,$$
$$c = \varphi_1^* (G_{uu}^0 \varphi_0 \varphi_0 + 2G_{u\lambda}^0 \varphi_0 + G_{\lambda\lambda}^0).$$

If now s is specified to denote arclength by demanding $2\xi_0^2 + \xi_1^2 = 1$ then the solution in the neighborhood of (u_0, λ_0) has the following expansion:

$$\begin{aligned} u(s) &= u_0 + (s - s_0)[\xi_0(s)\varphi_0 + \xi_1(s)\varphi_1] + (s - s_0)^2 v(s), \\ \lambda(s) &= \lambda_0 + (s - s_0)\xi_0(s) + (s - s_0)^2 \eta(s). \end{aligned} \tag{3.3}$$

A *pitchfork* bifurcation point ("B" in Figure 2.1), for example, is characterized by $a = 0$, $b \neq 0$, and thus

$$\xi_0(2b\xi_1 + c\xi_0) = 0. \tag{3.4}$$

Frequently these points are symmetry-breaking points (see **[23]** and the literature cited there); in this case $u_0 \dot{u}_0 = 0$ since $u_0 \in X_s$, $\varphi_1 \in X_a$, and $X = X_s \oplus X_a$, where X_s contains symmetric and X_a antisymmetric elements.

This short presentation should suffice to provide some background for the following.

4. Nonlinear elliptic eigenvalue problems. The problem in the case of equations will be of the form

$$G(u, \lambda) = 0, \qquad G\colon X^m \times \mathbf{R}^p \to X^m. \tag{4.1}$$

The solution may thus have m components equivalent to (4.1) being a system of m equations while there is in addition a dependence on p real parameters. These may enter in various ways. The coefficients or right-hand sides of the differential equations in (4.1) may depend on them, the boundary conditions, or the domain.

The following version of (4.1) underlies the program PLTMG **[1]**. Much of the following is joint work with R. Bank.

$$\begin{aligned} -\operatorname{div} a(x, y, u, \nabla u, \lambda) + f(x, y, u, \nabla u, \lambda) = 0 &\qquad \text{in } \Omega, \\ u = g_1(x, y, \lambda) &\qquad \text{on } \partial\Omega_1, \\ a \cdot n = g_2(x, y, u, \lambda) &\qquad \text{on } \partial\Omega_2 = \partial\Omega \backslash \partial\Omega_1, \end{aligned} \tag{4.2}$$

where $X \subset H^1(\Omega)$, $\Omega \subset \mathbf{R}^2$, $m = 1$, and a is a two-component vector function.

Since the solutions of (4.1) form a manifold in $X^m \times \mathbf{R}^p$, reductions in the dimensions m and p have to be made and additionally functionals of the solution have to be introduced to obtain graphical impressions of (parts of) $G^{-1}(0)$. We have decided to vary one of the components λ_j of λ at a time and to depict a functional $r_i = r_i(\lambda_j)$ of the solution. These functionals have the form

$$r_i(u) = \int_\Omega p_i(x, y, u, \nabla u, \lambda)\, dx + \int_{\partial\Omega} q_i(x, y, u, \nabla u, \lambda)\, ds. \tag{4.3}$$

The class of problems (4.2) is general enough to include many interesting applications as, for example, reaction-diffusion problems, but also homotopy continuation problems for which the parameters would be artificially introduced. A typical system (4.1) with equations of the form (4.2) is the nonlinear coupled system of VLSI device simulation

$$\begin{aligned} -\Delta u + e^{u-v} - e^{w-u} - k_1 &= 0, \\ \operatorname{div}(\mu_n e^{u-v} \nabla v) + k_2 &= 0, \\ -\operatorname{div}(\mu_p e^{w-u} \nabla w) + k_2 &= 0, \end{aligned} \tag{4.4}$$

for which numerical results were presented in [**4**].

5. Symmetry-breaking bifurcation. In this section we return briefly to the special but important class of bifurcation points mentioned in Section 4 at which symmetries are broken. These points are important mainly for two reasons. First, symmetries present in a problem in general give rise to bifurcations that similar problems without these symmetries do not have. Second, the discretizations of the problems that preserve some or all of their symmetries in general also possess the corresponding bifurcation points and these are not perturbed as in (2.5).

For simplicity let us consider the problem (2.1). We assume that a finite group of $\gamma \subset L(X)$ exists such that

$$G(Tu, \lambda) = TG(u, \lambda), \qquad \text{for all } T \in \gamma,\ u \in X,\ \lambda \in \mathbf{R}. \tag{5.1}$$

$L(X)$ denotes here the linear transformations on X. The subspace

$$X_s = \{u \in X,\ u = Tu \quad \text{for all } T \in \gamma\}$$

contains γ-symmetric elements and is invariant under $G(\cdot, \lambda)$. A typical case is $\gamma = \{I, S\}$, where I is the identity and $S \in L(X)$ such that $S^2 = I$, $S \neq I$.

In this case the space can be split as

$$X = X_s \oplus X_a \tag{5.2}$$

where

$$X_s = \{x \in S;\ Sx = x\} \quad \text{and} \quad X_a = \{x \in X, Sx = -x\}.$$

A special continuation method was proposed and analyzed for this simpler case in **[23]**, while an extension of this algorithm to more complicated groups γ was considered in **[27]**. In both cases numerical results for typical applications were given. General purpose software, namely the program PLTMG, to be discussed later, was applied to examples with symmetry-breaking bifurcation in **[25]**, **[26]**.

From (3.3), (3.4) it follows that for the case (5.2) the local representation of the primary and secondary (symmetry is broken) branches is given by

$$\begin{aligned} &u_1(s) = u_0 + w_1(s), && \lambda_1(s) = \lambda_0 + s, \\ &u_2(s) = u_0 + s\varphi_1 + w_2(s), && \lambda_2(s) = \lambda_0 + O(s^2), \\ &\|w_i(s)\| = O(|s|^i), \quad i = 1, 2, && w_1(s) \in X_s \end{aligned} \tag{5.3}$$

where without loss of generality s_0 was taken to be zero.

6. Continuation—two different approaches. If the parameter s introduced in (3.1) is taken as the arclength along the curve, i.e.,

$$\|\dot{u}_0\|^2 + \dot{\lambda}_0^2 = 1 \tag{6.1}$$

is used as a normalization then the following natural approximation of (6.1),

$$N_\sigma(u, \lambda) = \dot{u}_0(u - u_0) + \dot{\lambda}_0(\lambda - \lambda_0) - \delta\sigma = 0 \tag{6.2}$$

leads to the classical pseudo-arclength continuation method as described and analyzed, for example, in **[10]**, **[14]**.

From the known solution point (u_0, λ_0) of (4.1) predict the

$$(u_p, \lambda_p) = (u_0, \lambda_0) + \delta\sigma(\dot{u}_0, \dot{\lambda}_0) \tag{6.3}$$

that satisfies (6.2). Then correct by solving the augmented system

$$\begin{pmatrix} G(u, \lambda) \\ N(u, \lambda) \end{pmatrix} = 0 \tag{6.4}$$

iteratively starting with (u_p, λ_p). Each solution of (6.4) lies on the line perpendicular to the tangent to the branch in (u_0, λ_0) through (u_p, λ_p).

After various earlier works **[3]**, **[6]**, **[20]**, **[22]**, **[24]** joint efforts resulted in a similar but in several aspects quite different continuation method that was implemented in **[1]** and analyzed as well as applied in **[25]**, **[26]**, **[29]**.

Instead of the normalization (6.2) we consider

$$N_r(u,\lambda) = \theta \dot{r}_0(r - r_0) + (2-\theta)\dot{\lambda}_0(\lambda - \lambda_0) - \delta\sigma = 0 \tag{6.5}$$

where $0 \le \theta \le 2$ and r is a functional of u as in (4.3). $(\dot{u}_0, \dot{\lambda}_0)$ satisfies (6.1). Then the predicted point is found in the form

$$(u_p, \lambda_p) = (\gamma u_0 + \beta \dot{u}_0, \lambda_0 + \alpha \dot{\lambda}_0) \tag{6.6}$$

where (α, β, γ) are obtained from a least-squares solution of

$$\begin{pmatrix} N(u_p, \lambda_p) \\ (u_0, G(u_p, \lambda_p)) \\ (\dot{u}_0, G(u_p, \lambda_p)) \end{pmatrix} = 0. \tag{6.7}$$

Here we have assumed that X is a Hilbert space with scalar product $(\cdot, \cdot)$. The corrector iteration then consists of damped Newton for the analogous augmented system (6.4) starting with (u_p, λ_p). For a more detailed discussion of this method we refer to the above listed papers. The possible improvements over the classical pseudo-arclength method are apparent from the first example in **[26]**.

We mention here only that (6.5) suggests that continuation is done using target values in r ($\theta = 2$) and/or λ ($\theta = 0$) and that $\theta = 1$ corresponds to a pseudo-arclength method in (r, λ)-space. Since the r_i and λ_j are physically relevant functionals, respectively parameters, the method allows us to work constantly in terms of quantities of interest and eliminates artificial parameters as much as possible. It lends itself very much to an interactive implementation.

7. Implementation. As mentioned earlier the efforts of several researchers that had developed methods and programs for the efficient solution of nonlinear, parameter-dependent boundary value problems for elliptic partial differential equations were combined and an existing program was rewritten for parameter-dependent problems with (6.4)–(6.7) as continuation method **[1]**.

Single differential equations of the form (4.2) are discretized by linear finite elements. Continuation in r and/or λ target values is done on a coarse discretization of the problem. At any point adaptive local mesh refinement may be used to obtain via multigrid more accurate solutions with the same r and/or λ values, respectively orthogonal to the branch on the coarsest grid ($\theta = 1$). Multigrid is used only in a modular way, as an efficient method to solve linear subproblems.

The program detects and locates singular points; we refer to **[3]** for details. If, for example, a simple bifurcation point is located then one of the solutions of (3.2) is known. A finite difference approximation of the second derivatives in the second equation of (3.1) is used to obtain the other solution and branch switching may be accomplished.

In case the bifurcation point is of the symmetry-breaking type (5.3) then the choice of the functional $r = \|u\|$ that is used in **[1]** leads to a degeneracy. We note that in (u_0, λ_0) both $\dot{\lambda}_0$ and $\dot{r}_0$ are zero on the symmetry-breaking branch. The first follows from (5.3), the second from

$$\dot{r}_0 = (u_0, \dot{u}_0)/r_0$$

and $\varphi_1 \in X_a$, $u_0 \in X_s$. The condition (6.5) thus has to be modified or replaced in or very close to (u_0, λ_0).

While we refer again to the papers **[25]**, **[26]** for further results obtained by the program we list a few features that have been implemented recently. The user may now define several functionals r_i as in (4.3) and switch at any time between these as well as between the various parameters λ_j. For an example of the latter we refer to the second application in **[26]**. The classical multigrid technique used in **[1]** has now been replaced by hierarchical bases multigrid and a careful implementation of the streamline diffusion method in conjunction with the adaptive refinement to treat non-selfadjointness.

While the continuation strategy presented in Section 6 could lead to significant improvements for typical single reaction-diffusion equations, there is no straightforward generalization apparent with similar advantages in the case of systems of equations ($m > 1$ in (4.1)). In the following section some results will be reported for a system and the continuation technique implemented most recently will be outlined.

8. Application to VLSI-simplified prediction. In addition to various scalar equations the coupled system (4.4) of VLSI device simulation was solved using the above continuation method as implemented in PLTMG. The program was rewritten for this purpose to solve systems of the form (4.1), (4.2) of dimension 3 as described in **[4]**. As a test example a typical MOSFET device was chosen; see Figure 8.1.

For a complete description of the problem we refer to **[4]**. All voltages were kept constant except the drain voltage λ which was increased from 0 to 4 volts. The following table lists the required corrector iterations including those done before the steplength had to be cut back to assure convergence. Results are given for λ-continuation ($\theta = 0$ in (6.5)) as well as for r-continuation ($\theta = 2$) to the same points, where $r = (\|u\|^2 + \|v\|^2 + \|w\|^2)^{1/2}$. λ-continuation is done with λ-steps of .25 or .5 volts

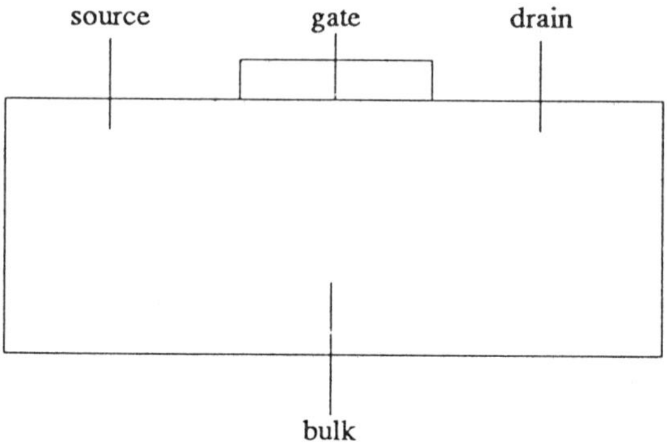

Figure 8.1. Schematic diagram of MOSFET

while the results for r-continuation are only given for the latter case. It is clear that r-continuation is somewhat less expensive, in particular for larger steps; but the computations showed that a straightforward generalization of the scalar program to systems in which the components in general represent physically different quantities with differing scales etc. does in general not lead to such a substantial improvement that the additional cost for the predictor, cf. (6.6) and (6.7), is justified.

		λ-continuation		r-continuation
λ	r	$\delta\lambda = .25$	$\delta\lambda = .5$	$\delta\lambda = .5$
0	1.95	21	21	14
.25	2.13	5	5	1
.75	2.84	27	40	24
1.25	3.83	34	37	29
1.75	4.95	23	27	25
2.25	6.16	12	15	16
2.75	7.42	15	19	10
3.25	8.72	15	25	11
4	10.7	6	9	9
TOTAL		158	198	139

Table 8.1. Continuation results for MOSFET

In the meantime, continuation results have been obtained for VLSI devices (CMOS inverter) for which the I-V curves, V voltage and I current, i.e., a functional of the form (4.3), exhibit turning points due to latch-up phenomena; see [**2**].

Various techniques have been proposed for the step-picking procedure once a continuation method has been chosen. We refer here only to **[11]**, **[30]** and the literature cited there. In order to make an established theory (namely that of damping Newton's method) available for continuation, the following simplified approach has recently been implemented in PLTMG.

We consider continuation from the known solution (u_0, λ_0) and define the residual

$$\text{res}(u, \lambda) = (\|G(u, \lambda)\|^2 + N_r^2(u, \lambda))^{1/2} \tag{8.1}$$

of the augmented system (6.4). We further assume that a steplength $\delta\sigma$ in (6.5) is known, for example as a consequence of a given r and/or λ target value. The point (u_0, λ_0) satisfies $\text{res}(u_0, \lambda_0) = \delta\sigma$.

The predicted point of the form

$$(u_p, \lambda_p) = (u_0 + q\dot{u}_0, \lambda_0 + q\dot{\lambda}_0)$$

is now found in such a way that its residual $\text{res}(u_p, \lambda_p)$ has sufficiently decreased in the sense of the damping strategy from the value $\delta\sigma$. This predictor requires only evaluations of G and N and has thus a minimal cost per iteration. Since the point predicted is along the tangent it will in general not allow one to use extraordinarily large steps as were possible for (6.6).

9. Variational inequalities. If the differential equation in (4.2) is self-adjoint the boundary value problem may equivalently be given in variational form, either as a minimum problem for a suitable functional (energy) or in weak form. We now consider problems which most easily are described variationally while on the level of differential equations they lead to a somewhat more complicated form involving a free boundary.

While much work has been done on variational inequalities that do not depend on a parameter including their efficient solution by multigrid and other techniques (see, for example, **[13]**, **[19]**), only recently have parameter-dependent variational inequalities received some attention. Theoretical continuation results have been obtained in **[7]**, **[17]** while methods have been proposed and applied in **[8]**, **[9]**, **[17]**, **[22]**, **[28]**.

On the product $X \times \mathbf{R}$, X a suitable Hilbert space, we look for solutions $(u, \lambda) \in K \times \mathbf{R}$ of the *variational inequality*

$$f'(u)(v - u) \geq \lambda g'(u)(v - u), \qquad \text{for all } v \in K. \tag{9.1}$$

Here f', g' are the first Gâteaux derivatives of continuous real functionals f and g defined on X, and K is a closed convex subset of X.

With the exception of a technical hypothesis that is verified in **[17]** for an important application we quote the main results of this paper. Defining the functional

$$I_\lambda(v) = f(v) - \lambda g(v) \tag{9.2}$$

and investigating when pairs (u, λ) are strict local minima of (9.2), we arrive at results about the continuation of solutions near (u, λ). Under technical assumptions (one of them being that $\lambda < \Lambda$, where Λ is the smallest eigenvalue of a certain linear variational inequality) there are solutions

$$(u(\varepsilon), \lambda(\varepsilon)) = (u + \varepsilon u_1 + o(\varepsilon), \lambda + \varepsilon)$$

to problem (9.1) for small enough $|\varepsilon|$ near the solution (u, λ). Here u_1 is the unique solution of

$$I''_\lambda(u)(u_1, v - u_1) \geq \operatorname{sign}(\varepsilon) g'(u)(v - u_1).$$

We remark that this result holds only for regular points along the solution branch of the variational inequality. It does not hold in fold (turning) points but may be generalized to transition points, i.e., points where the problem changes from one that can be described as a boundary value problem or nonlinear equation to one that has to be written as a free boundary problem or variational inequality.

In [7] theoretical results were given for problems including turning and transition points but for a very special form of (9.1) only. Basically the variational inequalities considered are

$$a(u, v - u) \geq \lambda(F(u), v - u), \qquad \text{for all } v \in K, \tag{9.3}$$

where $a(u, v) = \int_\Omega a_{ij}(x) u_{x_i} v_{x_j}\, dx$ and F is a positive increasing function.

10. Continuation for obstacle problems. While the numerical solution of parameter-dependent variational inequalities has only recently begun we report here about one interesting phenomenon first observed in **[28]**. The special form

$$\int_\Omega \nabla u \nabla (v - u)\, dx \geq \lambda \int_\Omega e^u (v - u)\, dx, \forall v \in K, \tag{10.1}$$

where Ω is a two-dimensional domain and

$$K = \{u \in H_0^1(\Omega), \quad u \leq \psi \text{ a.e. in } \Omega\} \tag{10.2}$$

was solved numerically in **[8]**, **[9]**, **[28]**. Here H_0^1 is the usual Sobolev space including boundary conditions. Before numerical results are reported we sketch the continuation method proposed in **[28]**.

We assume that (9.1) has been discretized, so that $X = \mathbf{R}^n$, $n = n(h)$, h the discretization parameter, and for the closed convex subset we take

$$K = \{u \in \mathbf{R}^n, u \leq d\}$$

where inequalities between vectors are to be understood componentwise. From a given pair (u_0, λ_0), $u_0 \in K$, with $\|u_0\| = r$ we now attempt, in the spirit of the continuation method in Section 6, to continue to a solution u with $\|u\| = r_t \neq r$.

The method will iteratively generate a sequence $\{u_k\}$, $u_k \in K$, that converges to the desired solution. For each k we define the set of active constraints by

$$I_k = I(u_k) = \{i \in \{1, \dots, n\}, u_{ki} = d_i\}.$$

Two matrices P_k and Q_k are defined by

$$P_k = (e_i)_{i \in I_k}, \qquad Q_k = E_n - P_k P_k^{\mathrm{T}}$$

where $e_i \in \mathbf{R}^n$ is the ith unit vector and E_n the $n \times n$ identity matrix. We further use the notation F_k, G_k for the Hessian matrices of f, g in (9.1) in u_k. The following method is a feasible direction method, as opposed, for example, to methods that incorporate the constraints through penalty terms.

The continuation method.

PREDICTOR. Let $u_0, \lambda_0, \dot{u}_0, \dot{\lambda}_0$ be given. Compute $\overline{\delta s}$ from $\|u_0 + \overline{\delta s}\dot{u}_0\| = r_t$ and δs from

$$\delta s := \min\left\{\overline{\delta s}, \min_{i \notin I_k}((d_i - u_{0i})/\dot{u}_{0i},\ \dot{u}_{0i} > 0)\right\}.$$

Set $u_1 = u_0 + \delta s \dot{u}_0$, $\lambda_1 = \lambda_0 + \delta s \dot{\lambda}_0$, $I_1 = I(u_1)$.

CORRECTOR. Set $k = 1$.

(i) Compute $q_k = \nabla f(u_k) - \lambda_k \nabla g(u_k)$ and terminate the iteration if $P_k^{\mathrm{T}} q_k \leq 0$, $\|Q_k q_k\| = 0$.

(ii) Set $|q_{kj}| := \max\{|q_{kj}|, (P_k^{\mathrm{T}} q_k)_i > 0\}$. If possible, deactivate the jth constraint, $\widetilde{I}_k = I_k - \{j\}$, and determine $\widetilde{Q}_k$; otherwise let $\widetilde{I}_k = I_k$, $\widetilde{Q}_k = Q_k$.

(iii) Solve the linear system

$$\begin{bmatrix} F_k - \lambda_k G_k & -\nabla g(u_k) \\ u_k^{\mathrm{T}} & 0 \end{bmatrix} \begin{bmatrix} \delta u_k \\ \delta \lambda_k \end{bmatrix} \begin{bmatrix} \lambda_k \nabla g(u_k) - \nabla f(u_k) \\ \frac{1}{2}(r_i^2 - u_k^{\mathrm{T}} u_k) \end{bmatrix}$$

but in the "free" variables ($i \notin \widetilde{I}_k$) only.

(iv) Determine the maximal admissible steplength $\overline{\alpha}_k$ from

$$\overline{\alpha}_k = \min_{i \notin \widetilde{I}_k}((d_i - u_{ki})/\delta u_{ki},\ \delta u_{ki} > 0)$$

and set $\alpha_k = \min(\overline{\alpha}_k, 1)$, $u_{k+1} = u_k + \alpha_k \delta u_k$, $\lambda_{k+1} = \lambda_k + \alpha_k \delta \lambda_k$. Set $k = k + 1$ and go to (i).

11. Spurious transition points. The problem (10.1), (10.2) was solved using the method given in Section 10. For the detailed results we refer to **[28]**. We have picked, for simplicity, a constant obstacle function ψ, such that $\psi > \|u_0\|_\infty$, where u_0 is the unique fold point this problem has. The following has to be expected for the bifurcation diagram in the (r, λ)-plane, where $r = \|u\|$ is the discrete L^2-norm. Beyond the turning point the solution will contact the obstacle ψ for the first time. There will be a nonsmooth transition point when this happens. For increasing r the curve must ultimately be flat, when $u_i = d_i$, $i = 1, \dots, n$. Schematically the curve had the following shape.

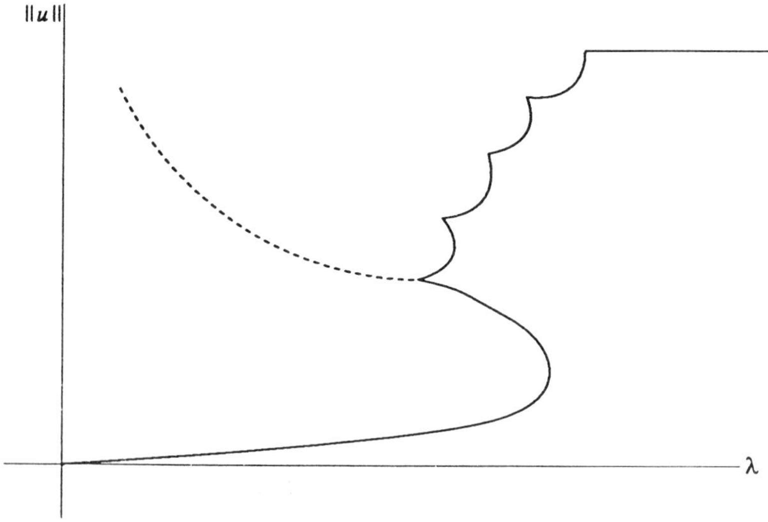

Figure 11.1. Schematic diagram for problem (10.1), (10.2)

While no mathematical theory explains this situation yet, a plausibility argument can easily be made for the occurrence of several spurious transition points.

After the solution has contacted the obstacle for the first time, there is a finite range of larger r-values for which no further contact occurs. In this range the problem in $\Omega \backslash B$, B the contact region, is an unconstrained problem of the Bratu type, the solution curve of which will asymptotically approach the r-axis for $r \to +\infty$. It may, therefore, turn around and generate left-turning fold points until the next constraints become active, meas(B) is increased. This is clearly a spurious effect, present only in the discretized problem and not in the continuous version.

12. Summary. In this paper we surveyed some recent work on continuation methods for both parameter-dependent nonlinear boundary value problems and variational inequalities. The first class of problems includes many important systems of elliptic differential equations, in particular the system for VLSI device simulation. The second class, elliptic variational inequalities, goes beyond the classical examples in continuum mechanics and includes, for example, problems in mathematical economics (quasivariational inequalities) for which the obstacle (cf. (10.2)) may even depend on the solution.

For the first class of problems, when the differential equation has the form (4.2), cooperation led to a sophisticated public domain program. Versions of this program have in particular also been used to compute I-V curves for VLSI devices including devices that exhibit latchup. This work is in progress and further fine-tuning is to be expected. Also, other classes of problems will be considered.

The continuation for variational inequalities is in a much earlier stage. Several algorithms have been proposed and successfully applied to some model problems. Theoretical results about continuation with respect to parameters have been obtained. In both these areas, however, much remains to be done. Just as parameter-dependent boundary value problems may exhibit spurious solutions, certain variational inequalities give rise to spurious effects of their discretizations.

BIBLIOGRAPHY

1. R. E. Bank, PLTMG *User's Guide*, Edition 4.0, Tech. Rep., Dept. Math., Univ of California, San Diego, 1985.

2. R. E. Bank, W. M. Coughran, Jr., W. Fichtner, E. H. Grosse, D. Rose, and R. K. Smith, *Transient simulation of silicon devices and circuits*, Numerical Analysis Manuscript 85-8, AT&T Bell Laboratories, Murray Hill, NJ 07974.

3. R. E. Bank and T. F. Chan, PLTMGC: *A multi-grid continuation program for parametrized nonlinear elliptic systems*, SIAM J. Sci. Statist. Comput. 7 (1986), 540–559.

4. R. E. Bank and H. D. Mittelmann, *Continuation and multi-grid for nonlinear elliptic systems*, Multigrid Methods II (W. Hackbusch and U. Trottenberg, eds.), Springer Lect. Notes in Math., **1228**, 1986.

5. J. H. Bolstad and H. B. Keller, *A multigrid continuation method for elliptic problems with folds*, SIAM J. Sci. Statist. Comput. 7 (1986), 1081–1104.

6. T. F. Chan and H. B. Keller, *Arc-length continuation and multi-grid techniques for nonlinear elliptic eigenvalue problems*, SIAM J. Sci. Statist. Comput. **3** (1982), 173–194.

7. F. Conrad, F. Issard-Roch, C. M. Brauner, and B. Nicolaenko, *Nonlinear eigenvalue problems in elliptic variational inequalities: A local study*, Comm. Partial Differential Equations **10** (1985), 151–190.

8. F. Conrad and R. Herbin, *Approximation of nonlinear eigenvalue problems associated with elliptic variational inequalities by continuation methods*, Innovative Numerical Analysis (C. A. Brebbia, ed.), Springer-Verlag, New York, 1986.

9. F. Conrad, R. Herbin, and H. D. Mittelmann, *Approximation of obstacle problems by continuation methods*, SIAM J. Numer. Anal. **25** (1988), 1409–1431.

10. D. W. Decker and H. B. Keller, *Path following near bifurcation*, Comm. Pure Appl. Math. **34** (1981), 149–175.

11. C. Den Heijer and W. C. Rheinboldt, *On steplength algorithms for a class of continuation methods*, SIAM J. Numer. Anal. **18** (1981), 925–947.

12. W. Hackbusch, *Multi-grid solution of continuation problems*, Iterative solution of nonlinear systems (R. Ansorge, T. Meis, and W. Törnig, eds.), Springer Lect. Notes in Math., **953**, 1982.

13. W. Hackbusch and H. D. Mittelmann, *On multi-grid methods for variational inequalities*, Numer. Math. **42** (1983), 65–76.

14. H. B. Keller, *Numerical solution of bifurcation and nonlinear eigenvalue problems*, Applications of Bifurcation Theory (P. Rabinowitz, ed.), Academic Press, New York, 1977.

15. T. Küpper, H. D. Mittelmann, and H. Weber (eds.), *Methods for bifurcation problems*, ISNM 70, Birkhäuser-Verlag, Basel, 1984.

16. E. Miersemann and H. D. Mittelmann, *A free boundary problem and stability for the nonlinear beam*, Math. Methods Appl. Sci. **8** (1986), 516–532.

17. ———, *On the continuation for variational inequalities depending on an eigenvalue parameter*, Math. Methods Appl. Sci. **11** (1989), 95–104.

18. H. D. Mittelmann and H. Weber, *Numerical methods for bifurcation problems—A survey and classification*, Bifurcation problems and their numerical solution (H. D. Mittelmann and H. Weber, eds.), ISNM 54, Birkhäuser-Verlag, Basel, 1980.

19. H. D. Mittelmann, *On the efficient solution of nonlinear finite element equations II. Bound-constrained problems*, Numer. Math. **36** (1981), 375–387.

20. ———, *Multigrid methods for simple bifurcation problems*, in Multigrid Methods (W. Hackbusch and U. Trottenberg, eds.), Springer Lect. Notes in Math., **960**, 1982.

21. ———, *Bifurcation problems for discrete variational inequalities*, Math. Methods Appl. Sci. **4** (1982), 243–258.

22. ———, *An efficient algorithm for bifurcation problems of variational inequalities*, Math. Comp. **41** (1983), 473–485.

23. ———, *Continuation near symmetry-breaking bifurcation points*, Numerical methods for bifurcation problems (T. Küpper, H. D. Mittelmann, and H. Weber, eds.), ISNM 70, Birkhäuser-Verlag, Basel, 1984.

24. H. D. Mittelmann and H. Weber, *Multi-grid solution of bifurcation problems*, SIAM J. Sci. Statist. Comput. **6** (1985), 49–60.

25. H. D. Mittelmann, *Multi-level continuation techniques for nonlinear boundary value problems with parameter-dependence*, Appl. Math. Comput. **19** (1985), 265–282.

26. ———, *A pseudo-arclength continuation method for nonlinear eigenvalue problems*, SIAM J. Numer. Anal **23** (1986), 1007–1016.

27. H. D. Mittelmann and B. Thomson, *An algorithm that exploits symmetries in bifurcation problems*, Numerical Techniques in Continuum Mechanics (W. Hackbusch and K. Witsch, eds.), Notes on Numerical Fluid Mechanics, **16**, Vieweg, Braunschweig, 1987.

28. H. D. Mittelmann, *On continuation for variational inequalities*, SIAM J. Numer. Anal **24** (1987), 1374–1381.

29. ———, *Multi-grid continuation and spurious solutions for nonlinear boundary value problems*, Rocky Mountain J. Math. **18** (1988), 387–401.

30. W. C. Rheinboldt and J. V. Burkhardt, *A locally parametrized continuation process*, ACM Trans. Math. Software **9** (1983), 215–325.

ARIZONA STATE UNIVERSITY, TEMPE, ARIZONA 85287-1804

Lectures in Applied Mathematics
Volume **25** (1990)

Recent Developments in the Numerical Solution of Differential/Algebraic Systems

LINDA R. PETZOLD

Abstract. In this paper we survey some recent developments in the numerical solution of nonlinear differential/algebraic equation (DAE) systems of the form $0 = F(t, y, y')$, where the initial values of y are known and $\partial F/\partial y'$ may be singular. These systems arise in the simulation of electrical networks, as well as in many other applications. DAE systems include standard form ODEs as a special case, but they also include problems which are in many ways quite different from ODEs. We examine the classification of DAE systems according to the degree of singularity of the system, and present some results on the analytical structure of these systems. We give convergence results for backward differentiation formulas applied to DAEs and examine some of the software issues involved in the numerical solution of DAEs. One-step methods are potentially advantageous for solving DAE systems with frequent discontinuities. However, recent results indicate that there is a reduction in the order of accuracy of many implicit Runge-Kutta methods even for simple DAE systems. We examine the current state of solving DAE systems by implicit Runge-Kutta methods. Finding a consistent set of initial conditions is often a problem for DAEs arising in applications. We explore some numerical methods for obtaining a consistent set of initial conditions.

1. Introduction. In this paper we survey some recent results on the solution by numerical ODE methods of linear and nonlinear differential/algebraic (DAE) systems of the form

$$0 = F(t, y, y'), \tag{1.1}$$

1980 *Mathematics Subject Classification* (1985 *Revision*). Primary 65L05.
This work was supported in part by the Applied Mathematical Sciences subprogram of the Office of Energy Research, U. S. Department of Energy by Lawrence Livermore National Laboratory under Contract No. W-7405-Eng-48.

where $\frac{\partial F}{\partial y'}$ may be singular. Differential/algebraic systems include standard form ODEs as a special case, but they also include problems which are in many ways quite different from ODEs. Initial value problems in DAEs arise in a wide variety of applications, including circuit and control theory, chemical kinetics, modelling of constrained mechanical systems, fluid dynamics, and robotics. We explore some classes of initial value problems which can be solved by backward differentiation formulas, and discuss some results on the order of convergence of implicit Runge-Kutta methods applied to DAE systems. Finding a consistent set of initial conditions is often a problem for DAE initial value problems arising in applications. We outline some recent work on a general algorithm for finding consistent initial conditions.

The basic idea of using a numerical method for solving DAE systems was introduced by Gear **[9]**, and consists of replacing y and/or y' in (1.1) by a difference approximation, and then solving the resulting equation for an approximation to y. In this paper we consider both backward differentiation formulas (BDF) and implicit Runge-Kutta methods (IRK) for defining the difference approximation. The simplest example of a numerical ODE method for (1.1) is the backward Euler method. Using this approach, the derivative $y'(t_n)$ at time t_n is approximated by a backward difference of $y(t)$ and the resulting system of nonlinear equations is solved for y_n,

$$0 = F\left(t_n, y_n, \frac{y_n - y_{n-1}}{h}\right), \tag{1.2}$$

where $h = t_n - t_{n-1}$. In this way the solution is advanced from time t_{n-1} to t_n.

This scheme of approximating the solution and/or the derivative in (1.1) and solving the resulting nonlinear equations is quite convenient for the solution of many different problems arising from practical applications. Systems in applications often are written naturally in the form (1.1). If we consider numerical methods for solving these very general systems, there will be no need to rewrite the system to fit into some special form. In addition, even though some DAE systems can be rewritten as standard form ODEs ($y' = f(t, y)$) or in other very special forms, often this is very inconvenient and it sometimes leads to a loss of sparsity in the matrices which are involved in the solution process. Finally, it is quite easy to explain to users of codes how to pose a problem in the form (1.1).

The disadvantage of this quite general formulation is that there are problems which can be written in the form (1.1) which are not solvable by ODE methods. Also there are problems which are solvable by some numerical ODE methods, but we must take very special care in choosing an appropriate method and also in implementing it. To distinguish these classes of problems requires a means of investigating the underlying structure of the DAEs.

2. Structure of DAE systems. In some sense the simplest class of DAEs is the class of linear constant-coefficient systems of the form

$$Ey' + Fy = g(t). \tag{2.1}$$

When $E + \lambda F$ is singular for all values of λ, then no solutions exist, or infinitely many solutions exist. We are not interested in the solution of these systems. The remaining systems, where $E + \lambda F$ is nonsingular, can be completely understood by transforming the system to *Kronecker canonical form* [**21**], [**8**]. These systems are *solvable* in the sense that solutions exist for all sufficiently smooth input functions $g(t)$, and two solutions which have the same initial value are identical. For solvable systems, there exist nonsingular matrices $\mathbf{P}$ and $\mathbf{Q}$ which decouple the system,

$$\begin{aligned} y_1'(t) + Cy_1(t) &= g_1(t), \\ Ny_2'(t) + y_2(t) &= g_2(t), \end{aligned} \tag{2.2}$$

where

$$\mathbf{Q}^{-1}y(t) = \begin{pmatrix} y_1(t) \\ y_2(t) \end{pmatrix}, \qquad \mathbf{P}g(t) = \begin{pmatrix} g_1(t) \\ g_2(t) \end{pmatrix}$$

into a "differential" part and a "singular" part. The matrix E is a block diagonal matrix whose blocks have the form

$$\begin{pmatrix} 0 & & & & \\ 1 & 0 & & & \\ & 1 & \ddots & & \\ & & \ddots & 0 & \\ & & & 1 & 0 \end{pmatrix}.$$

The most important concept in classifying these systems is the *index*. The index is the dimension of the largest block of E. The index is zero if there is no "singular" part to the system (that is, if the system is a standard form ODE or a system which can be rewritten as a standard form ODE by multiplying by E^{-1}). In general, the higher the index the more difficulties we are likely to encounter in trying to solve the system by a numerical ODE method.

To put these definitions into perspective, consider a canonical index-two problem

$$y_1' = g(t),$$
$$y_2 = y_1',$$

and note that for this system, in contrast to an ODE, the initial values must be consistent with the input function $g(t)$, and the solution is not as smooth as the input.

For nonlinear systems (1.1), we can define the *local index* as the index of the local constant-coefficient system at any given time, or the *global index*, which we will be primarily concerned with here, in terms of the number of differentiations of the system which are necessary to be able to solve for y' uniquely in terms of y and t [**6**].

DEFINITION 2.1. The global index of a solvable DAE (1.1) is the smallest nonnegative integer m such that F has m continuous derivatives and the nonlinear system

$$\begin{aligned} F(t, y, y') &= 0, \\ \frac{dF}{dt}(t, y, y', y'') = \frac{\partial F}{\partial y} y' + \frac{\partial F}{\partial y'} y'' + \frac{\partial F}{\partial t} &= 0, \\ &\vdots \\ \frac{d^m F}{dt^m}(t, y, y', y'', \ldots, y^{(m+1)}) &= 0 \end{aligned} \tag{2.3}$$

can be solved for y' uniquely in terms of y and t: $y' = \phi(y, t)$.

There are two important classes of index-one systems. The first is the general *fully implicit* index-one system (1.1). If we assume that the rank of $\partial F / \partial y'$ is constant, and that the index is identically equal to one in a neighborhood of the solution, then we will refer to these systems as *uniform index one*. The second class of index-one systems is that of *semi-explicit* index-one systems. These are systems which are written in the special form

$$\begin{aligned} 0 &= F_1(x, x', y, t), \\ 0 &= F_2(x, y, t), \end{aligned} \tag{2.4}$$

where $\partial F_1 / \partial x'$ is nonsingular. The system (2.4) is index one iff $\partial F_2 / \partial y$ is nonsingular. These semi-explicit index-one systems arise frequently in applications. It is important to distinguish them from the general index-one system (1.1) because there are numerical methods which perform well for (2.4) but poorly for (1.1).

For high-index systems in the general form (1.1), numerical ODE methods such as the backward Euler method can be unstable for small stepsizes [**12**]. It is often possible to reduce the index of a semi-explicit system by analytically differentiating the constraints [**12**]. Campbell [**6**] has devised an algorithm for solving higher-index systems which repeatedly differentiates the entire system and solves the resulting overdetermincd system for y' in terms of y and t in the spirit of Definition 2.1. However, in its present form this algorithm is not useful for the solution of most problems in applications because it requires knowledge of the analytic partial derivative matrices of F and time derivatives of these matrices and requires a large degree of smoothness which may not be present in all applications.

Despite the discouraging results on the instability of numerical ODE methods for general high-index linear systems, we can identify important subsets of high-index nonlinear problems for which we can obtain positive results. These systems arise in the modelling of electrical networks and constrained mechanical systems and in the solution of the equations of fluid flow. For index two, these are semi-explicit DAE systems of the form (2.4). The system is index two if $(\partial F_2/\partial x)(\partial F_1/\partial y)$ is nonsingular. For index-three systems, these are semi-explicit systems which can be written in a *triangular form* which includes the class of index-three constrained mechanical systems described in the next section.

Gear [**10**] has recently noted a simple relationship between the fully implicit index-one system and the semi-explicit index-two system with which it is often possible to transfer methods and convergence results from one class of problems to the other.

3. Backward differentiation formulas. In this section we explore some results on the solution by BDF methods of nonlinear index-one systems and some special index-two and index-three nonlinear systems which occur frequently in practical applications. We find that BDF methods converge with the expected order of accuracy for several large classes of DAEs of practical interest.

To solve (1.1) by a BDF method, we replace $y'(t_n)$ by a k-step backward differentiation formula,

$$\rho y_n = \sum_{i=0}^{k} \alpha_i y_{n-i}, \tag{3.1}$$

to obtain the system of nonlinear equations

$$F\left(t_n, y_n, \frac{\rho y_n}{h}\right) = 0. \tag{3.2}$$

Then we have the following result for BDF applied to index-one systems **[12]**.

THEOREM 3.1. *If F is uniform index one and is differentiable with respect to y and y', the solution of* (1.1) *by the k-step BDF method with fixed stepsize h for $k < 7$ converges to order $O(h^k)$ if all initial values are correct to order $O(h^k)$.*

An extension of this theorem to variable stepsizes is given in **[11]**.

While the BDF methods converge as expected for index-one problems, there are still practical difficulties in implementing these methods for this class of problems **[17]**, **[19]**. One of the difficulties is to obtain a consistent set of initial conditions. We will discuss this problem in section 5. It should be noted that systems whose index is less than or equal to one are the problems that general-purpose codes **[17]** are designed to handle.

For higher-index systems, we have noted earlier that it is not possible to obtain convergence even of backward Euler, in general. However, we can obtain some encouraging results **[16]** if we restrict our attention to semi-explicit higher-index systems (2.4). Systems of this form arise for example in the simulation of electrical networks, the solution of constrained mechanical systems of the form

$$\begin{aligned} M(q)q'' &= f(q, q', t) + G(q)\lambda, \\ \phi(q) = 0, &\qquad \frac{\partial \phi}{\partial q} = G^{\mathrm{T}}, \end{aligned} \tag{3.3}$$

where q are the positions and λ the Lagrange multipliers, and the solution of the finite difference or finite element approximation to the equations of fluid flow

$$\begin{aligned} &M\dot{U} + (K + N(U))U + CP = f(U, P), \\ &C^{\mathrm{T}} U = 0, \end{aligned} \tag{3.4}$$

where U is the approximation to the velocity and P approximates the pressure. The index of the mechanical systems (3.3) is three and of the fluid flow equations (3.4) is two.

A simple example of a system of form (3.3) is given by the equations describing a pendulum in Cartesian coordinates. Let L denote the length of the bar, λ the force in the bar, and x and y the coordinates of the infinitesimal ball of mass one located at the free end of the bar. Then x, y, and λ solve the DAE system:

$$\begin{aligned} x'' &= \lambda x, \\ y'' &= \lambda y - g, \\ 0 &= \frac{1}{2}(x^2 + y^2 - L^2), \end{aligned}$$

where g is the gravity constant. The index of this system is three.

For semi-explicit index-two systems and for index-three constrained mechanical systems of the form (3.3), the k-step constant-stepsize BDF method converges to order of accuracy $O(h^k)$ if the initial values are sufficiently accurate [**16**], [**2**]. Gear et al. [**11**] generalized this result to show that variable stepsize BDF methods converge for semi-explicit index-two systems.

It is possible to use a general purpose code based on backward differentiation formulas to solve these special high-index nonlinear systems [**19**]. However, there are some practical difficulties which must be dealt with. It can be shown that for an index m system, the iteration matrix which the code uses in the Newton iteration for solving the nonlinear equation (3.2) has a condition number which is $O(1/h^m)$. This difficulty can be remedied by scaling the equations and the variables. The convergence test and error test must also be modified to allow a variable-stepsize BDF code to solve these types of problems.

4. Implicit Runge-Kutta methods. In this section we describe some results on the properties of implicit Runge-Kutta methods applied to index-one DAE systems. These methods are potentially advantageous over multistep methods for some systems (for example, for systems with frequent discontinuities). However, some care must be taken in choosing an implicit Runge-Kutta method which is appropriate for DAEs, as these methods do not in general attain the same order of accuracy for DAEs as they do for ODEs.

An M-stage implicit Runge-Kutta (IRK) method applied to the system of DAEs (1.1) is written as

$$(4.1)\qquad \begin{aligned} &F\left(t_{n-1}+c_i h, y_{n-1}+h\sum_{j=1}^{M} a_{ij}Y_j', Y_i'\right)=0,\\ &y_n = y_{n-1}+h\sum_{i=1}^{M} b_i Y_i'. \end{aligned}$$

The method can be written in the shorthand notation which displays the matrix of coefficients,

$$\begin{array}{c|cccc} c_1 & a_{11} & a_{12} & \cdots & a_{1M}\\ c_2 & a_{21} & a_{22} & \cdots & a_{2M}\\ \vdots & \vdots & \vdots & \ddots & \vdots\\ c_M & a_{M1} & a_{M2} & \cdots & a_{MM}\\ \hline & b_1 & b_2 & \cdots & b_M \end{array}$$

We will assume that the matrix $\mathscr{A} = (a_{ij})$ is nonsingular.

We will first give some results on the properties of implicit Runge-Kutta methods applied to different classes of index-one systems, and then we will discuss the properties of these methods applied to nonlinear semi-explicit index-two systems. First we need to define some terminology.

The Runge-Kutta method will be called *strictly stable* if the difference between a perturbed Runge-Kutta step

$$(4.2)\qquad \begin{aligned} &F\left(t_{n-1}+c_i h, z_{n-1}+h\sum_{j=1}^{M} a_{ij}Z_j' + \delta_n^{(i)}, Z_i'\right)=0, \qquad i=1,2,\dots,M,\\ &z_n = z_{n-1}+h\sum_{i=1}^{M} b_i Z_i' + \delta_n^{(M+1)}, \end{aligned}$$

where $z_0 = y_0 + \delta_0^{(M+1)}$, $\|\delta_n^{(i)}\| \le \Delta$, and an unperturbed Runge-Kutta step (4.1) satisfies $\|y_n - z_n\| \le K_0\Delta$, where $0 < h \le h_0$ and K_0, h_0 are constants depending only on the method and the DAE.

Defining the *stability constant* $r = 1 - b^{\mathrm{T}}\mathscr{A}^{-1}\varepsilon_M$, where $\varepsilon_M = (1, 1, \dots, 1)^{\mathrm{T}}$, it is easy to show for index-one DAEs that the IRK method (4.1) is stable iff the method coefficients satisfy the *strict stability condition* $|r| < 1$. This stability condition for DAEs is related to the stability criterion $|R(z)| \le 1$, $R(z) = 1 + zb^{\mathrm{T}}(I - z\mathscr{A})^{-1}\varepsilon_M$, where $z = h\lambda$, for the stiff model problem $y' = \lambda y$ because $\lim_{|z|\to\infty} R(z) = 1 - b^{\mathrm{T}}\mathscr{A}^{-1}\varepsilon_M$.

Applying the IRK method (4.1) to a canonical constant-coefficient index-one system $u = g(t)$, we find that the method is locally accurate to order $O(h^{k_a+1})$ iff the method coefficients satisfy $A_1(k_a)$, where $A_1(w)$ is defined by

$$(4.3) \qquad A_1(w):\ b^{\mathrm{T}}\mathscr{A}^{-1}c^j = 1, \qquad j = 1, 2, \ldots, w,$$

where $c^j = (c_1^j, c_2^j, \ldots, c_M^j)^{\mathrm{T}}$. If the method satisfies the strict stability condition and the order of the IRK method for purely differential systems is k_d, then it is easy to show that the method is globally accurate to order $O(h^{\min(k_d, k_a+1)})$ for index-one constant-coefficient DAE systems.

One class of IRK methods which appears to be particularly promising for the solution of DAEs is that of *stiffly accurate* methods. These are methods whose coefficient matrices satisfy $c_M = 1$, $a_{M_j} = b_j$, $j = 1, \ldots, M$ and $\mathscr{A}$ is nonsingular. For semi-explicit index-one systems, these methods have the property which is analogous to BDF that the constraint equations are satisfied exactly at the end of each time step. Griepentrog [**13**] and Deuflhard et al. [**7**] have shown that there is no order reduction for stiffly accurate methods applied to semi-explicit index-one systems.

For more general implicit Runge-Kutta methods, Roche [**20**] has recently derived a set of order conditions which include (4.3) using the theory of Butcher series and rooted trees, for semi-explicit index-one systems.

The IRK methods are, in general, even less accurate for fully implicit nonlinear index-one systems than for semi-explicit systems. The additional loss of accuracy comes about because of mixing which can occur between the errors in the differential and singular parts of the system. To state the results, we must first define some terminology.

Defining

$$C(w):\ \sum_{j=1}^{M} a_{ij}c_j^{k-1} = c_i^k/k, \qquad i = 1, 2, \ldots, M, \qquad k = 1, 2, \ldots, w,$$

$$(4.4) \qquad B(w):\ \sum_{j=1}^{M} b_j c_j^{k-1} = 1/k, \qquad k = 1, 2, \ldots, w,$$

then we have the following result, which is shown in [**4**].

THEOREM 4.1. *Suppose that* (1.1) *is uniform index one and linear in* y', *the Runge-Kutta method satisfies the stability condition* $|r| \le 1$, *the errors in the initial conditions are* $O(h^G)$ *and the errors in terminating the Newton iterations are* $O(h^{G+\delta})$, *where* $\delta = 1$ *if* $|r| = 1$ *and* $\delta = 0$ *otherwise, and* $G \geq 2$. *Then the global errors satisfy* $\|e_n\| = O(h^G)$ *where*

$$G = \begin{cases} q, & \textit{if } C(q) \textit{ and } B(q), \\ q+1, & \textit{if } C(q), B(q+1) \textit{ and } -1 \leq r < 1. \end{cases}$$

It should be noted that implicit Runge-Kutta methods with $|r| = 1$ can be unstable for some fully implicit index-one systems [**1**].

It should also be noted that Theorem 4.1 gives only a lower bound on the order. Thus it is possible that the observed order for some methods and some problems may exceed this lower bound. In numerical experiments described in [**18**], [**4**], we found that the lower bounds were actually observed for some methods in the stiff ODE literature. A few methods appear to perform better than the lower bounds would indicate.

These results for index one DAEs are in agreement with the order reduction results for stiff ODEs given in [**5**] in the limit as the stiffness becomes infinite [**4**].

For nonlinear semi-explicit index-two systems, because of the close relationship between fully implicit index-one problems and semi-explicit index-two systems, we have that the global error in the "differential" variable x is given by Theorem 4.1. The global errors in the y variable are $O(h^{G_y})$ [**3**], where G_y is given by

$$G_y = \begin{cases} q & \text{if } C(q), B(q), A_1(q), \\ q+1 & \text{if } C(q), B(q+1), A_1(q+1), A_2(q+1), \end{cases}$$

where A_2 are the order conditions for Runge-Kutta methods applied to canonical constant-coefficient index-two systems and are given by

$$\begin{aligned} A_2(w)\colon\ & b^{\mathrm{T}}\mathscr{A}^{-1}\varepsilon_M = b^{\mathrm{T}}\mathscr{A}^{-2}c, \\ & b^{\mathrm{T}}\mathscr{A}^{-2}c^i = i,\ i = 2, 3, \ldots, w. \end{aligned}$$

Numerical experiments given in [**3**] show that the additional order reduction does occur for index-two systems, including problems which are of interest in applications. It is difficult to find higher-order IRK methods which maintain the same order of accuracy in all of the variables for an index-two system.

5. Determining consistent initial conditions. Given the available software for solving DAEs with BDF methods [**17**], [**14**], often the most difficult part of solving a problem in applications is to determine a consistent set of initial conditions with which to start the computation.

More precisely, we formulate this problem as follows. Given information about the initial state of the system which is sufficient (in a mathematical sense) to specify a unique solution to the DAE, determine the complete initial state $(x(t_0), x'(t_0))$ corresponding to this unique solution. For example, the user may specify information about the solution and/or its derivatives at the initial time, and the problem is to determine the remaining values of the solution and its derivatives. This work is described in more detail in Leimkuhler et al. **[15]**.

For a simple algorithm, one might consider substituting $y = y_0$, $y' = y_0'$ and solving the resulting system, together with the user-defined information, for y_0 and y_0'. For example, for an ODE $y' = Ay$ together with user-defined information $\mathbf{C}y_0' + \mathbf{D}y_0 = q_0$ at the initial point, we have

$$\begin{pmatrix} -\mathbf{A} & \mathbf{I} \\ \mathbf{D} & \mathbf{C} \end{pmatrix} \begin{pmatrix} y_0 \\ y_0' \end{pmatrix} = \begin{pmatrix} 0 \\ q_0 \end{pmatrix}, \tag{5.1}$$

which we can solve uniquely for y_0 and y_0' provided the left-hand matrix is of full rank and q_0 is in the range of the matrix $\mathbf{CA} + \mathbf{D}$. In this case, these are the same conditions which the matrices $\mathbf{D}$ and $\mathbf{C}$ must satisfy to ensure that the user has given enough information about the initial state to specify a unique solution to the mathematical problem.

On the other hand, consider the index-one DAE given by

$$\begin{aligned} Y_1' + Y_2' + Y_1 &= g_1(t), \\ Y_2 &= g_2(t). \end{aligned} \tag{5.2}$$

To specify a unique solution to the DAE, it is sufficient to give the value of either Y_1 or Y_1' at t_0, because Y_2 and Y_2' must satisfy the constraint and its derivative. However, evaluating Y_1, Y_1', Y_2, and Y_2' at time t_0 in (5.2), it is apparent that it is not possible to obtain $Y_1(t_0)$ uniquely if only $Y_1'(t_0)$ is specified. Thus this simple algorithm fails to give the solution, although the user has given enough information to specify a unique solution to the original problem.

Clearly the difficulty with the above procedure for equations (5.2) is that the simple algorithm has no way of obtaining the information about the derivative of the constraint which is inherent in the system. In the fully implicit index-one problem it is not in general possible to isolate the constraints and differentiate them. Thus we are led to consider the following algorithm, motivated by Definition 2.1.

Solve

$$F(t_0, y_0, y_0') = 0,$$

$$\frac{dF}{dt}(t_0, y_0, y_0', y_0'') = 0,$$

$$\vdots$$

$$\frac{d^mF}{dt^m}(t_0, y_0, y_0', \dots, y_0^{(m+1)}) = 0, \tag{5.3}$$

coupled with the user-defined information

$$B(y_0, y_0', t_0) = 0. \tag{5.4}$$

It is easy to show that if the user-defined information (5.4) is sufficient to determine a unique solution to the DAE, then (5.3), (5.4) have a solution $(y_0, y_0', \dots, y_0^{(m+1)})$ and the first two components y_0, y_0' are the solution to the DAE at t_0. Note that the higher derivatives of y are not determined uniquely by this algorithm.

Since it is obviously not practical to obtain the derivatives of F analytically, we are led to consider approximating the derivatives of F. Because F is possibly only defined for $t \geq t_0$, it is natural to consider using one-sided differences. For example, the simplest one-sided difference is given by

$$D_hF = \frac{F(y_0 + hy_0', y_0' + hy_0'', t_0 + h) - F(y_0, y_0', t_0)}{h}.$$

We can define higher-order difference approximations by

$$D_hF = \frac{1}{h}\left[\sum_{i=1}^{s} \alpha_i F(y_0 + hc_i y_0', y_0' + hc_i y_0'', t_0 + hc_i) - \left(\sum_{i=1}^{s} \alpha_i\right) F(y_0, y_0', t_0)\right] \tag{5.5}$$

by choosing the constants α_i, c_i appropriately. The higher derivatives of F can be approximated similarly.

The algorithm obtained by replacing the consistency equations (5.3) by their approximations (5.5), coupled with the user-defined information (5.4) produces a rank-deficient over-determined nonlinear system. Unlike the analytic consistency equations, the approximate system may not have an exact solution because some of the derivative approximations to F may be approximating user-defined information. Thus the approximate system is solved in a least squares sense. However, the solution of this system is complicated by the rank-deficiency of the Jacobian. Using the structure of the system, it is possible to show that the minimum norm solution to this nonlinear least squares problem converges to the correct solution as the approximations D_h become more accurate, and to formulate a scheme for replacing the system with a full-rank system which has the same solution for y_0, y_0'. Numerical results for this technique appear to be promising.

BIBLIOGRAPHY

1. U. Ascher, *Symmetric Schemes and Differential-Algebraic Equations*, SIAM J. Sci. Statist. Comput. **10** (1989), 937–949.
2. K. E. Brenan and B. Engquist, *Backward Difference Approximations of Nonlinear Differential-Algebraic Equations*, Math. Comp. **51** (1988), 659–676.
3. K. E. Brenan and L. R. Petzold, *The Numerical Solution of Higher Index Differential/Algebraic Equations by Implicit Runge-Kutta Methods*, SIAM J. Numer. Anal. **26** (1989), 976–996.
4. K. Burrage and L. R. Petzold, *On Order Reduction for Runge-Kutta Methods Applied to Differential/Algebraic Systems and to Stiff Systems of ODEs*, SIAM J. Numer. Anal. (to appear).
5. K. Burrage, W. H. Hundsdorfer and J. A. Verwer, *A Study of B-Convergence of Runge-Kutta Methods Applied to Stiff Systems*, Computing **36** (1986), 17–34.
6. S. L. Campbell, *A General Form for Solvable Linear Time Varying Singular Systems of Differential Equations*, SIAM J. Math. Anal. **18** (1987), 1101–1115.
7. P. Deuflhard, E. Hairer, and J. Zugck, *One-Step and Extrapolation Methods for Differential-Algebraic Systems*, Preprint Nr. 318, Universität Heidelberg, Institut für Angewandte Mathematik, 1985.
8. F. R. Gantmacher, *The Theory of Matrices*, vol. 2, Chelsea, New York, 1964.
9. C. W. Gear, *Simultaneous Numerical Solution of Differential/Algebraic Equations*, IEEE Trans. Circuit Theory **CT-18** (1971), 89–95.
10. C. W. Gear, *Differential-Algebraic Equation Index Transformations*, SIAM J. Sci. Statist. Comput. **9** (1988), 39–47.
11. C. W. Gear, B. Leimkuhler, and G. K. Gupta, *Automatic Integration of Euler-Lagrange Equations with Constraints*, J. Comput. Appl. Math. **12** and **13** (1985), 77–90.
12. C. W. Gear and L. R. Petzold, *ODE Methods for the Solution of Differential/Algebraic Systems*, SIAM J. Numer. Anal. **21** (1984), 367–384.
13. E. Griepentrog, *Initial Value Problems for Implicit Differential Equations Systems*, Humboldt-Universität zu Berlin, reprint, 1984.
14. A. C. Hindmarsh, *ODE Solvers for use with the Method of Lines*, Advances in Computer Methods for Partial Differential Equations, IV, R. Vichnevetsky and R. S. Stepleman, eds., IMACS, New Brunswick, N.J., 1981, 312–316.
15. B. J. Leimkuhler, L. R. Petzold, and C. W. Gear, *On the Consistent Initialization of Differential-Algebraic Equations*, Lawrence Livermore National Laboratory, UCRL-97867, December 1987.
16. P. Lötstedt and L. Petzold, *Numerical Solution of Nonlinear Differential Equations with Algebraic Constraints I: Convergence Results for Backward Differentiation Formulas*, Math. Comp. **46** (1986), 491–516.
17. L. R. Petzold, *A Description of DASSL: A Differential/Algebraic System Solver*, IMACS Trans. Sci. Comput., Vol. 1, R. S. Stepleman, ed., 1982.

18. L. R. Petzold, *Order Results for Implicit Runge-Kutta Methods Applied to Differential/Algebraic Systems*, SIAM J. Numer. Anal. **23** (1986), 837–852.

19. L. Petzold and P. Lötstedt, *Numerical Solution of Nonlinear Differential Equations with Algebraic Constraints* II: *Practical Implications*, SIAM J. Sci. Statist. Comput. **7** (1986), 720–733.

20. M. Roche, *Runge-Kutta Methods for Differential Algebraic Equations*, SIAM J. Numer. Anal. **26** (1989), 963–975.

21. R. F. Sincovec, B. Dembart, M. A. Epton, A. M. Erisman, J. W. Manke, and E. L. Yip, *Solvability of Large-Scale Descriptor Systems*, Boeing Computer Services Company, Seattle, Washington, 1979.

COMPUTING AND MATHEMATICS RESEARCH DIVISION, L-316, LAWRENCE LIVERMORE NATIONAL LABORATORY, LIVERMORE, CALIFORNIA 94550